THIRD EDITION

CIVIL DRAFTING TECHNOLOGY

David A. Madsen

Department Chairperson
Drafting Technology
Past Member Board of Directors
American Design Drafting Association (ADDA)
Premier Autodesk Training Center
Softdesk Authorized Training Center for Civil Engineering
ESRI Authorized ArcVIEW Learning Center
Clackamas Community College, Oregon City, Oregon
ADDA Drafter Certified Curricula

Terence M. Shumaker

Manager, Premier Autodesk Training Center
Instructor, Drafting Technology
Clackamas Community College
Oregon City, Oregon

Prentice Hall
Upper Saddle River, New Jersey   Columbus, Ohio

Library of Congress Cataloging-in-Publication Data
Madsen, David A.
    Civil drafting technology / David A. Madsen, Terence M. Shumaker.
    —3rd ed.
        p.   cm.
    Includes index.
    ISBN 0-13-751181-7 (pbk.)
    1. Mechanical drawing.   I. Shumaker, Terence M.   II. Title.
T353.M196   1998
624'.022'1—dc21

97-9270
CIP

Cover art/photo: Background drawing: Courtesy of Otak, Inc.,
Lake Oswego, Oregon. Image of woman at
computer: © Mike Zens/Corbis. Image of
man surveying: © Michael Philip Manheim/
International Stock.
Editor: Ed Francis
Production Editor: Louise N. Sette
Design Coordinator: Julia Zonneveld Van Hook
Cover Designer: Russ Maselli
Production Manager: Deidra M. Schwartz
Marketing Manager: Danny Hoyt

This book was set in Century by Maryland Composition and was printed and
bound by R.R. Donnelley & Sons Company. The cover was printed by Phoenix
Color Corp.

 © 1998, 1994, 1983 by Prentice-Hall, Inc.
Simon & Schuster/A Viacom Company
Upper Saddle River, New Jersey 07458

Printed in the United States of America

10 9 8 7 6 5 4 3 2 1

ISBN: 0-13-751181-7

Prentice-Hall International (UK) Limited, *London*
Prentice-Hall of Australia Pty. Limited, *Sydney*
Prentice-Hall of Canada, Inc., *Toronto*
Prentice-Hall Hispanoamericana, S. A., *Mexico*
Prentice-Hall of India Private Limited, *New Delhi*
Prentice-Hall of Japan, Inc., *Tokyo*
Simon & Schuster Asia Pte. Ltd., *Singapore*
Editora Prentice-Hall do Brasil, Ltda., *Rio de Janeiro*

Preface

This text is intended to be a comprehensive instructional package in the area of civil drafting. The authors have used materials that have been tested in the classroom for several years plus input and ideas from industry, specifically civil engineering companies. Our aim is to provide the student or employee with a well-rounded view of the civil drafting field and the types of drawings and skills associated with that field.

The book is arranged in eleven chapters, each dealing with a specific subject area. We feel that the arrangement lends itself well to a one-term or semester course but contains enough information and problems to fit courses of varying length. Each chapter is followed by a test composed of short essay questions, fill-ins, true/false, and sketching. Problems in the form of drawings, which close each chapter, can be completed in the text, on separate sheets of paper, or using computer-aided drafting (CAD) as instructed. The tests and problems enable the student to directly apply the information presented in each chapter to realistic situations.

The field of civil drafting is one filled with variety and excitement. From surveying to construction, courthouse research to artistic interpretation, the opportunities offer many challenges. The authors have drawn on their experiences of surveying in the jungles of Georgia to building houses in the foothills of the Oregon Cascades; from designing wastewater piping to designing solar homes. And most importantly, we have drawn on our collective experiences in teaching the varied aspects of drafting at the community college level. Additionally, we have included the work of contributing professionals in the applications of CAD throughout the entire text, and the coverage of geographic information systems (GIS) as applied to civil drafting technology. Many educators who have successfully used the first and second editions of this text have provided comprehensive reviews. These reviews resulted in expanded coverage in every chapter, plus new and varied drafting problem assignments.

The use of this text in the prescribed manner imparts to the student a broad understanding of civil drafting and a working knowledge of the basic components of mapping. With this knowledge and skill a variety of job opportunities are open to the student, and with those opportunities, we sincerely hope, a challenging career. Keep in mind that mapping requires accuracy, neatness, and an eye for creative and uncluttered layout.

Our thanks must be extended to the Oregon Department of Education for allowing us to draft from some of their illustrations.

We hope that your experiences with civil drafting and mapping are just as exciting as ours. Good luck.

Special Thanks and Acknowledgments

We want to give special thanks to our contributing authors and the professionals who provided comprehensive reviews of the text and assisted with new and varied problem assignments:

Contributing Authors

Computer-aided drafting applications and third edition review: Cathy Stark, Engineering Technician II, CADD, City of Portland, Oregon.

Geographic Information Systems: Keith J. Massie, Geographic Information Systems Analyst, Metro (Portland, Oregon).

Reviewers

Earl Faulkner and Meredith Lambert, both of ITT Technical Institute.

David Madsen
Terence Shumaker

Contents

3

LOCATION AND DIRECTION *71*

4

MAPPING SCALES *105*

5

MAPPING SYMBOLS *119*

6

LEGAL DESCRIPTIONS AND PLOT PLANS *135*

7

CONTOUR LINES *165*

Introduction to Civil Drafting Technology

This chapter discusses maps in general and some of the different types in use today. Information about civil engineering companies, their map drafting requirements, and employment opportunities is also covered.

Topics to be discussed include:

- Characteristics of maps
- Types of maps
- Civil engineering companies
- Map requirements
- Schooling
- Cartography

MAPS IN GENERAL

Maps are defined as graphic representations of part of or the entire earth's surface drawn to scale on a plane surface. Constructed and natural features can be shown by lines, symbols, and colors. Maps have many different purposes depending on their intended usage. A map can accurately provide distances, locations, elevations, best routes, terrain features, and much more.

Some maps, such as aeronautical and nautical maps, are more commonly referred to as *charts*. This distinction is shown in the following discussion about types of maps.

Map Title Block and Legend

When you use or read a map the first place to look is the *title block* and *legend.* The information given here will tell immediately if you have the correct map. Other valuable information about map scales, symbols, compass direction, and special notes will also be given.

TYPES OF MAPS

Aeronautical Charts

Aeronautical charts are used as an aid to air travel. These charts indicate important features of land, such as mountains and outstanding landmarks (See Figure 1–1). Commonly prepared in color and with relief-shading methods, aeronautical charts are a very descriptive representation of a portion of the earth's surface.

FIGURE 1–1. A typical aeronautical chart. (Reproduced by permission of the National Ocean Survey [NOAA], U.S. Department of Commerce.)

Contour lines are often provided with 200- to 1000-ft intervals. There is a comprehensive amount of information regarding air routes, airport locations, types of air traffic, radio aids to navigation, and maximum elevation of features. Look at Figure 1–1 and you can see all of the detail shown in an aeronautical chart.

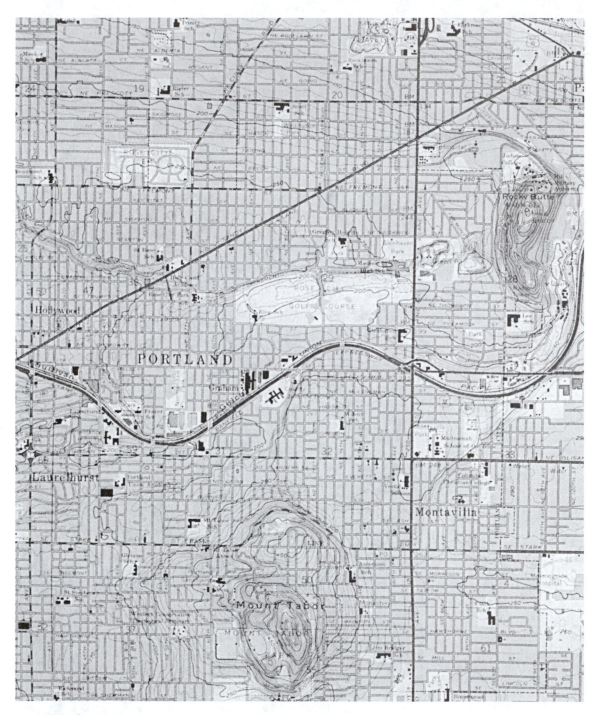

FIGURE 1–2. A typical cadastral map. (Reproduced by permission of the U.S. Geological Survey.)

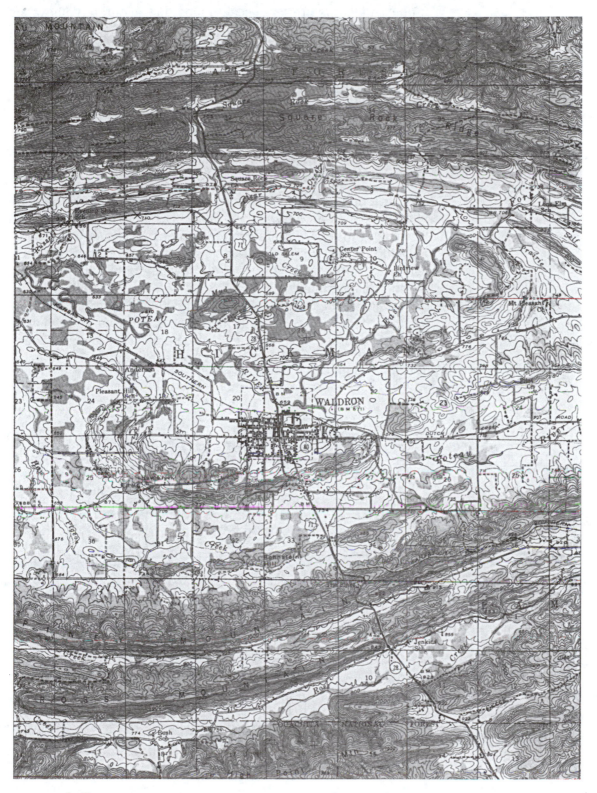

FIGURE 1–3. The quadrangle map shown here is a type of cadastral map.

Cadastral Maps

Cadastral maps are large-scale maps that accurately show the features in a city or town. These types of maps are often used for city development, operation, and taxation. Figure 1–2 provides an example of a cadastral map.

A *quadrangle map* is a type of cadastral map showing the division of land into grids known as sections. This type of map, shown in Figure 1–3, is used in the rectangular survey system discussed in Chapter 6.

Engineering Maps

Construction projects of all kinds are detailed to show the complete layout in an *engineering map*. The information provided includes the location and dimensions of all structures, roads, parking areas, drainage ways, sewers, and other utilities. Elevations of features and contour lines are optional. See Figure 1–4 for an example.

Engineering maps may also include plats. Plats are carefully detailed maps of construction projects such as subdivisions showing building lots. These may also be plot or site plans, which are plats of an individual construction site. Figures 1–5 and 1–6 provide examples.

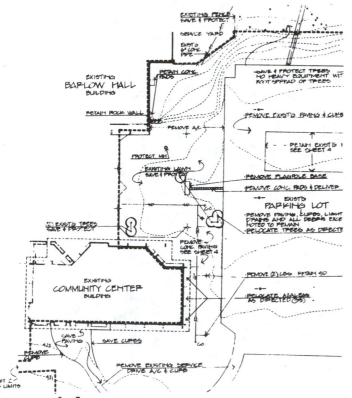
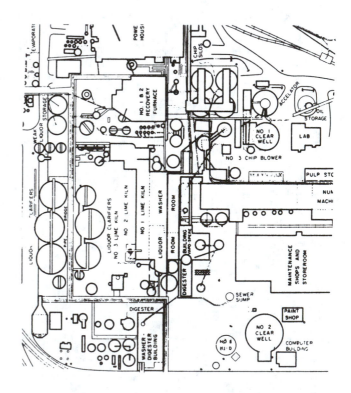

FIGURE 1–4. Typical engineering site plans.

EICHOLAN ESTATES

A PART OF THE NE 1/4 OF THE NW 1/4 OF SEC. 11
T.2S.,R.2E.,W.M.
CLACKAMAS COUNTY, OREGON

SHEET 1 OF 5 SHEETS

DATE: OCT 1993
PREPARED BY: TANZER & ASSOC.

	LOT	ANGLE	RADIUS	ARC LENGTH	CHORD	CHORD BEARING
BLK ①	1	90°	20.00'	31.42'	28.28'	N45°00'W
	3	90°	20.00'	31.42'	28.28'	N45°00'E
	6	90°	20.00'	31.42'	28.28'	N45°00'W
	7	90°	20.00'	31.42'	28.28'	S45°00'W
BLK ②	1	90°	20.00'	31.42'	28.28'	N45°00'E
	5	16°	70.00'	19.55'	19.48'	N8°00'E
	6	58°	70.00'	70.86'	68.87'	N45°00'E
	7	16°	70.00'	19.55'	19.48'	N82°00'E
	9	90°	20.00'	31.42'	28.28'	N45°00'W
	10	90°	20.00'	31.42'	28.28'	N45°00'W
	13	51°	20.00'	17.80'	17.22'	N84°30'W
	13	113°	50.00'	98.61'	83.39'	N83°30'W
BLK ③	1	113°	50.00'	98.61'	83.39'	N64°30'W
	2	51°	20.00'	17.80'	17.22'	N64°30'W
	3	90°	20.00'	31.42'	28.28'	N45°00'E
	4	45°	70.00'	54.98'	53.57'	N45°00'W
	5	45°	70.00'	54.98'	53.57'	N67°30'W
	7	51°	20.00'	17.80'	17.22'	N22°30'W
BLK ④	7	113°	50.00'	98.61'	83.39'	N25°30'W
	1	113°	50.00'	98.61'	83.39'	N6°30'E
	2	51°	20.00'	17.80'	17.22'	N6°30'W
BLK ⑤	2	45°	70.00'	54.98'	53.57'	N25°30'E
	3	90°	20.00'	31.42'	28.28'	S45°00'E
	4	90°	20.00'	31.42'	28.28'	S22°30'E
	6	90°	20.00'	31.42'	28.28'	N45°00'W
	①	90°	45.00'	70.68'	63.64'	N45°00'E
	②	90°	45.00'	70.68'	63.64'	N45°00'E
	③	90°	45.00'	70.68'	63.64'	N45°00'W
	④	90°	45.00'	70.68'	63.64'	N45°00'W

FIGURE 1-5. A typical subdivision with building lots.

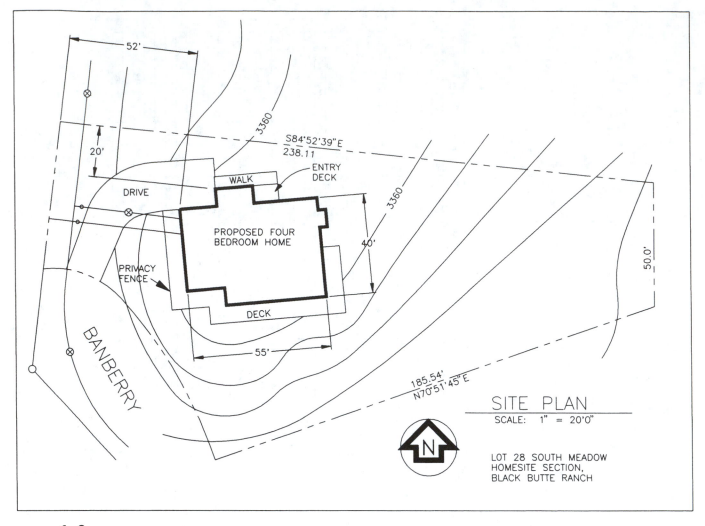

SITE PLAN
SCALE: 1" = 20'0"

LOT 28 SOUTH MEADOW
HOMESITE SECTION,
BLACK BUTTE RANCH

FIGURE 1-6. An individual plat with residential construction site.

Geographical Maps

Geographical maps are usually prepared at a small scale. These maps commonly show large areas of the earth, depicting continents, countries, cities, rivers, and other important features. A map of the world or maps of individual countries or states are considered geographical maps. Figure 1–7 represents a typical geographical map.

Hydrologic Maps

Hydrologic maps accurately show the hydrographic boundaries of major river basins. In the United States these maps are prepared by the U.S. Geological Survey in cooperation with the U.S. Water Resources Council. Hydrologic maps, which are used for water and land resource planning, are published at a scale of

FIGURE 1–7. A typical geographical map. (Reproduced by permission of the U.S. Geological Survey.)

1:500,000 (1 in. equals about 8 mi). The maps are printed in color and contain information on drainage, culture, and hydrographic boundaries. Figure 1–8 provides a sample hydrologic map.

Military Maps

Military maps can be any map that contains information of military importance or serves a military use. A military map may be used by a soldier in the field and may have information about ter-

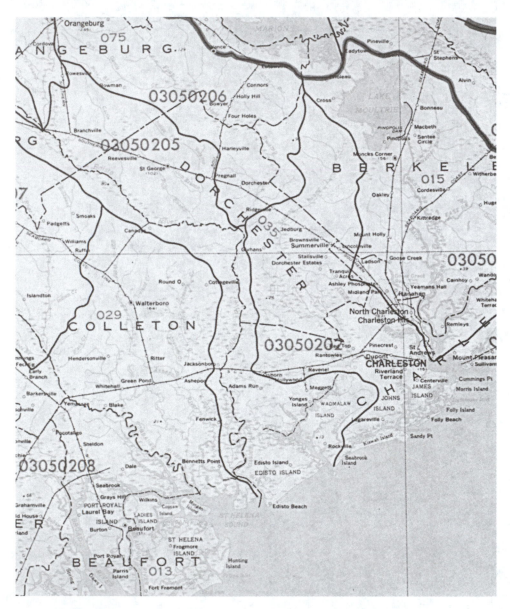

FIGURE 1–8. A typical hydrologic map. (Reproduced by permission of the U.S. Geological Survey.)

rain, concealment, and cover. It may also be a map of a large geographical area that can be used for military planning.

Nautical Charts

Nautical charts are special maps used as an aid to water navigation. These charts provide such information as water depths, clearances of bridges, and overhead cables. They also show navigation lanes, lighthouses, beacons, and buoys. Figure 1–9 shows a sample nautical chart.

FIGURE 1–9. A typical nautical chart. (Reproduced by permission of the National Ocean Survey [NOAA], U.S. Department of Commerce.)

Photogrammetric Maps

Aerial photographs are used to make *photogrammetric maps.* This process is the most widely used method of preparing maps. Aerial photos are taken at certain intervals and are controlled by stations on the earth's surface. These photos can be accurately scaled, and are easily read and transferred to paper using special stereoscopic instruments. Figure 1–10 shows an aerial photograph.

Compare the map in Figure 1–11 to the aerial photograph in Figure 1–10 and you can see how a map is created from an aerial photograph.

FIGURE 1–10. A typical photogrammetric map. (Reproduced by permission of the U.S. Geological Survey.)

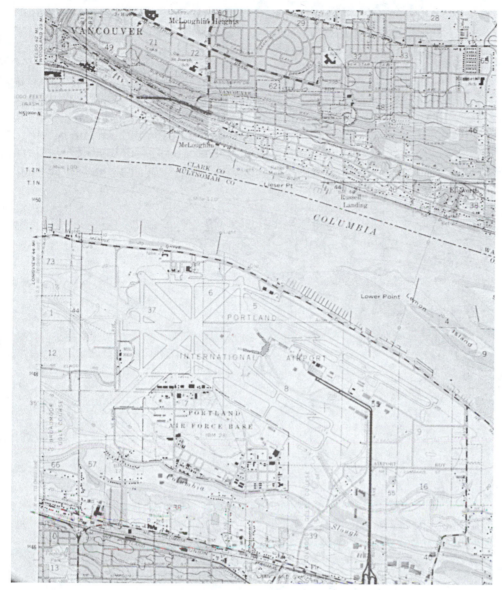

FIGURE 1–11. A map created from an aerial photograph. (Reproduced by permission of the U.S. Geological Survey.)

Topographic Maps

Topographic maps accurately show the shape of the earth by the use of contour lines. Contour lines represent all the points on the ground of equal elevation above sea level. The spacing of the contour lines is determined by the grade of the land. On very steep terrain the lines are close together, because changes in elevation come quickly. On terrain that slopes gradually the contour lines are farther apart, as it takes longer to reach a change in elevation. The contour lines are drawn at equal changes in elevation: for example, every 2, 5, or 10 ft. Usually, at least every fifth contour is broken along its length and the elevation is inserted. Contour lines are often drawn in brown. Topographic maps show streams, lakes, and rivers in blue. Woodland features are represented in green. Features that are constructed by people, such as buildings and roads, are often shown in black on a topographic map. Most of the types of maps previously discussed may often use the elements of a topographic map. Most construction projects involving shape, size, location, slope, or configuration of land can be aided by the use of topographic maps. There are probably thousands of varied uses for these maps. Figure 1–12 shows a topographic map.

FIGURE 1–12. A typical topographic map. (Reproduced by permission of the U.S. Geological Survey.)

Planning Maps

Maps used by planners may include a variety of those previously mentioned. A *planning map* may have a photogrammetric or topographic base map overlaid with appropriate information such as zoning boundaries, urban growth boundaries, or population density. This type of map may then be used with hydrologic maps or engineering maps to show existing conditions and to plan for improvements. Planning maps that use different "layers" or "levels" from digitally recorded and produced data (information that is gathered and utilized by computers) are relatively easy to compile and have a wide variety of uses. Computer-aided drafting (CAD) applications used in conjunction with a geographic information system (GIS) are discussed in Chapter 11. Depending on the specific need, planning maps are often produced in color and may be either large or small scale. Figure 1–13 shows a planning map.

Digital Terrain Model Maps

Digital terrain model (DTM) maps have been digitally recorded and produced utilizing a grid of elevation points. Different viewpoints of a site may be observed with the use of a CAD station, and the vertical scale may be magnified to display minute elevation changes at sites that are relatively flat. These maps are helpful in slope analysis and in determining hydrographic boundaries, as well as in the planning of highways, subdivisions, and under-

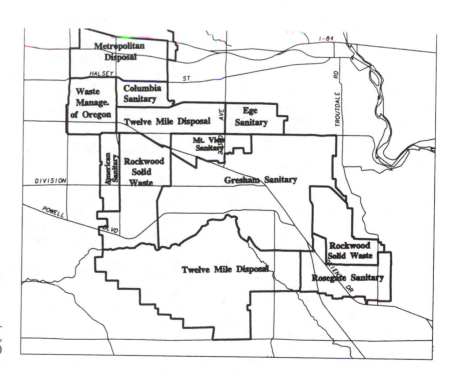

FIGURE 1–13. A typical planning map. (Reproduced by permission of Metro, Portland, Oregon.)

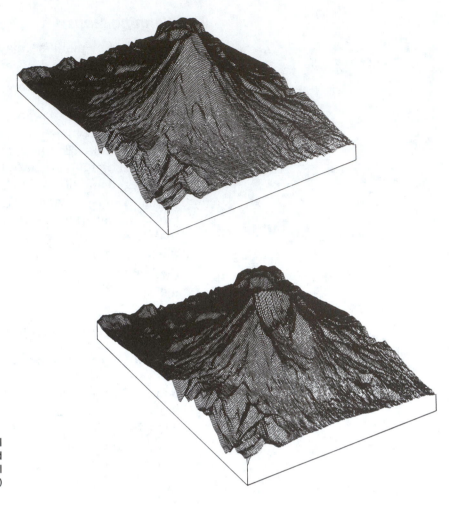

FIGURE 1–14. A digital terrain model (DTM) map of Mt. St. Helens. (Reproduced by permission of the U.S. Geological Survey.)

ground utilities that rely on gravity flow. Figure 1–14 represents a DTM of Mt. St. Helens in Washington state before and after the May 18, 1980 eruption.

CIVIL ENGINEERING COMPANIES

Civil engineering is concerned with the design of bridges, roads, dams, canals, and similar projects. Civil engineering companies are located nationwide in most cities. Some of these companies specialize in certain aspects of the industry, whereas others are quite diversified. The following is a list of some of the tasks that civil engineering companies may take part in:

- Land planning and subdivision
- Transportation

- Flood control
- Irrigation and drainage
- Sewage and water treatment
- Municipal improvements
- Environmental studies
- Land and construction surveys
- Construction inspection
- Refuse disposal
- Mapmaking
- Power plants
- Hydrologic studies
- Foundation work and soil analysis
- Agribusiness

A complete directory of consulting engineers is available from the American Consulting Engineers Council.

Drafting salaries of workers at civil engineering firms are usually competitive with those in other technologies. Working conditions vary but are usually excellent. Companies have a wide range of employee benefits. Check your local area regarding salary ranges and schooling requirements for entry-level drafters. Areas of the country differ in these concerns.

MAP REQUIREMENTS

Maps serve a multitude of purposes. Some maps may be used to show the construction site for a new home, while another map may show the geography of the world. Civil engineering companies primarily prepare maps that fall into the first category: construction site plans and maps relating to the civil projects previously described. The chapters covered in this workbook provide you with the basic information to continue a more in-depth study or on-the-job training.

The materials in most common use in civil drafting are CAD plots on vellum. Vellum provides a readily reproducible original at minimal cost. Companies that require a more durable original may use polyester film. Polyester film, commonly referred to by its trade name of Mylar™, reproduces better than vellum but is more costly. When polyester film is used, the drafter often uses liquid ink pens to create the CAD plot. A more detailed description of drafting materials can be found in most basic drafting texts.

CARTOGRAPHY

Cartography is the art of making maps and charts. A cartographer is a highly skilled professional who designs and draws maps. Cartography is considered an art. The cartographer is a master in the use of a variety of graphic media, mechanical lettering methods, and artistic illustration.

Civil drafting and cartography are quite similar in that both professions deal with the making of maps. However, civil drafting is generally concerned with maps and plans for construction and other civil-related projects. Cartography requires that the technician use more graphic skills in the preparation of printed documents and maps. Often the job title "cartographer" requires four years of education, with emphasis in civil engineering, geography, navigation, optics, geodesy, or cartography.

SCHOOLING

Technical schools and community colleges throughout the United States and Canada have drafting programs. Schools may provide a specific drafting education in mechanical, architectural, civil, piping, structural, technical illustration, sheet metal, electrical, or computer drafting. Other drafting programs may provide their students with a more general curriculum that may have courses in each of several of these areas. Often, the school focuses on the industrial needs of the immediate area. The best thing to do is to identify the school program that will best serve your specific goals. Civil drafting is offered in many technical schools and community colleges.

Your specific goals in civil drafting may include learning one of several available computer-aided drafting (CAD) software programs. Some technical schools and community colleges offer CAD classes in conjunction with drafting classes, and some places of employment provide CAD instruction to their employees who do drafting. While drafting with the aid of a computer is a skill in itself, your sound background and competence in essential drafting knowledge and technique are necessary to your success as a CAD operator. It is preferable that you combine manual drafting with your studies of CAD so that basic drafting skills are not overlooked. Therefore if you are interested in civil drafting as a career, whether or not you study CAD, your schooling should include the development of some of these fundamental skills:

- Drafting skills, line work, lettering, neatness, and the use of equipment

- Use of bearings and azimuths
- Use of the engineer's scale
- Scale conversion
- Drawing contour lines, and converting field notes
- Use of mapping symbols
- Preparation of a plat and interpretation of legal descriptions
- Development of plans and profiles
- Layout of highways, centerlines, curves, and delta angles
- Drawing cuts and fills
- Basic use and knowledge of surveying equipment
- Math through basic trigonometry

BASIC CIVIL DRAFTING TECHNIQUES

Line work in civil drafting is done using a number of optional techniques depending on the purpose of the map. For example, many engineering maps are drawn using pencil on vellum. This is a common inexpensive media for maps drawn for a variety of applications. Some maps, such as highway layouts, are typically drawn using ink on polyester film. Lines on maps should be dark, crisp, and easy to reproduce regardless of the method used to make the lines. Figure 1–15 shows a small variety of lines used in civil draft-

LINE TYPE	PEN NO	IN	MM
CITY, COUNTY OR STATE BOUNDARY	4	.047	1.40
SUBDIVISION BOUNDARY	4	.047	1.40
PROPERTY LINE AND LOT LINE	1	.020	0.50
EASEMENT LINE	0	.014	0.35
CENTER LINE	0	.014	0.35
PROPOSED WATER	2.5	.028	0.70
PROPOSED SEWER	3	.031	0.80
EXISTING WATER	0	.014	0.35
EXISTING SEWER	0	.014	0.35

FIGURE **1–15.** Lines commonly used on civil drafting.

FIGURE 1–16. Lettering typically used on civil drawings.

ing. Many more line types will be introduced throughout this text. The recommended line width is given in technical pen tip number, inches, and millimeters. Notice that the major boundary lines and proposed structures are drawn thicker than property lines and existing structures. Computer-aided drafting (CAD) makes drawing lines easy and provides consistently uniform line thicknesses.

The words on a drafting project are referred to as *lettering*, when done using manual methods, or as *text*, when computer-aided drafting is used. Although the style of lettering may vary depending on the type of project or the company making the drawing, it is generally a single-stroke letter form similar to the example shown in Figure 1–16. Use short single pencil strokes when doing freehand lettering. Lettering is often done with a soft 2H, H, or F lead in a 0.5 mm mechanical pencil. Always use horizontal guide lines to control the letter height. Some drafters also use vertical guide lines to help direct the vertical strokes. The guide lines should be drawn very lightly using a 4H, 6H, or nonreproducible blue lead.

The lettering on some civil drafting projects used in the construction industry is an architectural style. This type of lettering has a free form that is more artistic in nature than the traditional lettering previously discussed. This type of lettering is done completely freehand, or the vertical strokes may be done with the help of a drafting triangle or vertical drafting machine scale. This is done by placing a small triangle on the parallel bar or horizontal drafting machine scale. Make the vertical strokes along the triangle edge and then quickly make the other strokes freehand. A sample is shown in Figure 1–17.

Much lettering in the civil drafting industry has been done using mechanical lettering equipment. Mechanical lettering equipment is purchased in kits with templates, a scriber, and an assortment of technical pen widths. While practice is important to become good at using this equipment, experienced drafters can letter neatly and quickly with lettering equipment. Figure 1–18 shows lettering equipment being used.

The manual lettering methods previously discussed are

ABCDEFGHIJKLMNOPQR
STUVWXYZ 1234567890

FIGURE 1–17. Draw vertical lines to start letters with straight vertical strokes. Complete the rest of the letter using freehand methods.

FIGURE 1–18. Using mechanical lettering equipment.

rapidly being replaced by computer-aided drafting. The otherwise time-consuming task of lettering on a drawing becomes as easy as typing at a keyboard when CAD is used. The quality of CAD lettering is uniform and easy to read. CAD lettering is generally many times faster than manual lettering, and additional productivity is gained when changes are needed. Most computer-aided drafting systems make a variety of text styles and fonts available depending on the application, as shown in Figure 1–19. A *font* is all of the uppercase and lowercase letters and numerals of a particular letter face design.

ABCDEFGHIJKLMNOPQRSTUVWXYZ
1234567890

ABCDEFGHIJKLMNOPQRSTUVWXYZ
1234567890

ABCDEFGHIJKLMNOPQRSTUVWXYZ
1234567890

ABCDEFGHIJKLMNOPQRSTUVWXYZ
1234567890

FIGURE 1–19. Samples of CAD text fonts.

TEST

Part I

Define the following terms. Use your best lettering technique.

1-1 Map

1-2 Contour lines

1-3 Aerial photos

1-4 Plats

1-5 Civil drafting

Part II

Multiple choice: Circle the response that best describes each statement.

1-1 Small-scale maps that commonly show large areas of earth, depicting continents and countries are called:

a. Aeronautical charts
b. Cadastral maps
c. Geographical maps
d. Nautical charts

1-2 Maps that accurately show the shape of the earth by the use of contour lines are called:

    a. Photogrammetric maps
    b. Topographic maps
    c. Engineering maps
    d. Geographic maps

1-3 Maps that accurately show the boundaries of major river basins are called:

    a. Geographical maps
    b. Nautical charts
    c. Cadastral maps
    d. Hydrologic maps

1-4 Maps detailed to show the layout of a construction project are called:

    a. Geographical maps
    b. Cadastral maps
    c. Engineering maps
    d. Topographic maps

PROBLEMS

P1-1 Below each line on Figure P1–1, draw five more lines exactly the same using your drafting pencil and a straightedge. Make your lines as dark and crisp as you would on a real drawing.

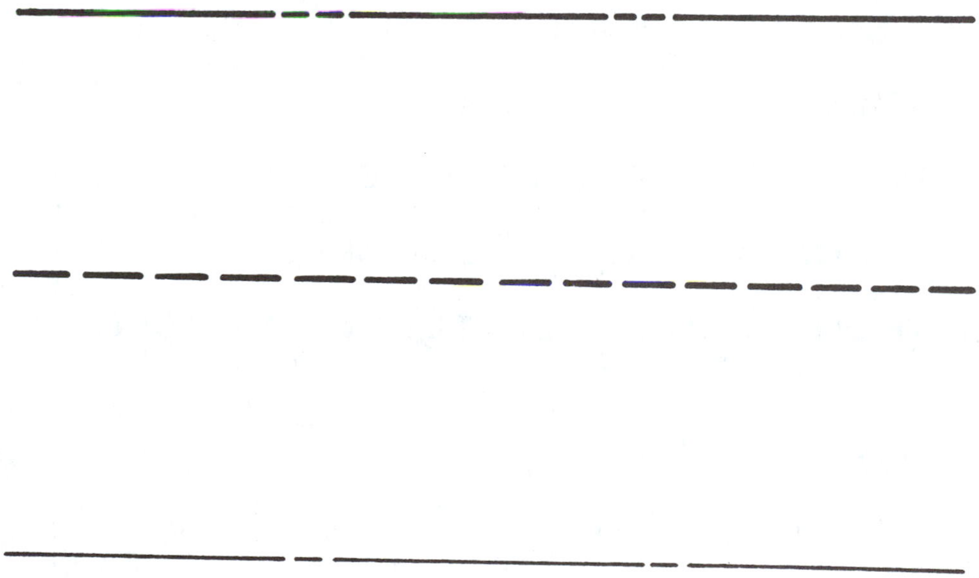

FIGURE **P1–1.**

P1–2 Using your best freehand lettering skill, duplicate each sentence written on Figure P1–2. Make your own very light guide lines 1/8 in. apart to guide your lettering. Try to make your lettering the same as the example unless otherwise indicated by your instructor. Use a soft 2H, H, or F pencil lead and a 0.5 mm automatic pencil.

THE QUALITY OF THE FREEHAND LETTERING GREATLY

AFFECTS THE APPEARANCE OF THE ENTIRE DRAWING.

MANY CIVIL DRAFTING TECHNICIANS USE FREEHAND

LETTERING IN PENCIL OR INK TO CREATE MAPS.

PROPER FREEHAND LETTERING IS DONE WITH A SOFT,

SLIGHTLY ROUNDED POINT DRAFTING PENCIL 2H, H, OR

F DEPENDING UPON THE INDIVIDUAL PRESSURE.   THE

LETTERING IS DONE BETWEEN VERY LIGHTLY DRAWN

GUIDE LINES.   THESE GUIDE LINES ARE DRAWN

PARALLEL SPACED EQUAL TO THE HEIGHTS OF THE

LETTERS.   GUIDE LINES HELP TO KEEP YOUR

LETTERING UNIFORM IN HEIGHT. LETTERING STYLES

MAY VARY BETWEEN COMPANIES, WHILE SOME

COMPANIES REQUIRE MECHANICAL LETTERING DEVICES.

FIGURE **P1–2.**

P1-3 Practice an architectural lettering style by duplicating each sentence lettered on Figure P1–3. Make your own very light guide lines 1/8 in. apart, to guide your lettering. Try to make your lettering the same as the example unless otherwise indicated by your instructor. Use a soft lead in a 0.5 mm automatic pencil.

ARCHITECTURAL LETTERING IS MORE

ARTISTIC IN NATURE THAN TRADITIONAL

LETTERING USED ON MOST DRAWINGS.

SOME DRAFTERS PREFER TO DRAW

THE VERTICAL STROKES USING THE

EDGE OF A TRIANGLE OR DRAFTING

MACHINE SCALE, WHILE OTHERS DO

FIGURE **P1–3.** ALL OF THEIR LETTERING FREE HAND.

P1-4 Use mechanical lettering equipment, if available, to duplicate each sentence lettered on Figure P1–4.

MECHANICAL LETTERING EQUIPMENT WITH

TECHNICAL INK PENS HAS BEEN COMMONLY

USED IN CIVIL DRAFTING TO PRODUCE

QUALITY LETTERING.   AS WITH MANY

DRAFTING SKILLS, IT TAKES SOME PRACTICE

TO BECOME PROFICIENT USING MECHANICAL

FIGURE **P1–4.** LETTERING EQUIPMENT.

P1-5 Use a technical ink pen, if available, to draw the lines displayed in Figure P1–1. Use the space provided below, and make the lines the same width and line type displayed in the example.

P1-6 Use a computer-aided drafting system, if available, to draw the lines displayed in Figure P1–1. Make the lines the same width and line type displayed in the example.

P1-7 Use a computer-aided drafting system, if available, to duplicate each line shown on Figure 1–15. Use text styles similar to the example, or substitute another text style if these are not available.

P1-8 Use a computer-aided drafting system, if available, to create two variations of both the mechanical and the architectural styles of text. Suggestions:

Fit the text into a compressed space.

Slant the text at a 15-degree angle.

Stretch the text without increasing the height.

P1-9 Classify the type of map shown on Figure P1–5. Describe how you came to your conclusion.

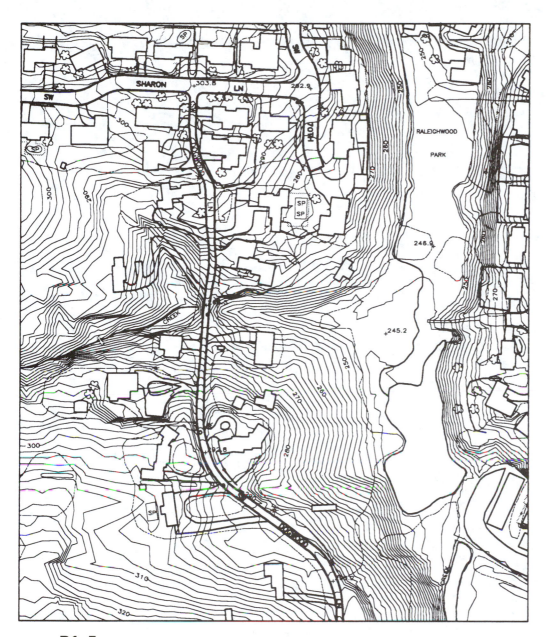

FIGURE P1–5. (Reproduced by permission of City of Portland, Oregon.)

Surveying Fundamentals

Accurate maps are created using information obtained from a variety of surveying methods. *Surveying* is the process used to obtain information about natural and human-made features. The ancient Egyptians employed people called *rope stretchers*, who used ropes marked at intervals with knots to measure and lay out plots and farm land. Today we can use satellites to locate positions on the earth, and sophisticated electronic surveying instruments to determine elevation, location, and distances. This chapter examines some of the methods and instruments used in gathering information from which maps are made.

Specifically, topics covered include:

- Types of surveys
- Land measurement techniques
- Elevation measurement
- Recording of measurements
- Traversing methods, including electronic traversing
- The Global Positioning System

THE SHAPE AND SIZE OF THE EARTH

The earth is referred to as a sphere but it is technically an *oblate spheroid,* or a sphere flattened at the ends (poles). The approximate measurements of the earth are:

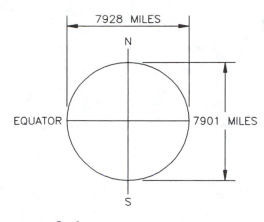

Equatorial diameter	7928 mi (12,756 km)
Equatorial circumference	24,907 mi (40,075 km)
Length of 1 degree of longitude at equator	69.186 mi
Polar diameter	7901 mi (12,713 km)
Polar circumference	24,822 mi (39,938 km)
Length of 1 degree of latitude	68.95 mi (approx.)

FIGURE 2–1. The diameter of the earth measured through the poles is approximately 27 ml less than the diameter measured at the equator.

The diameter of the earth measured through the poles is approximately 27 miles less than the diameter measured at the equator. See Figure 2–1. The rotation of the earth creates the 27-mile bulge at the equator. This causes a variation in the length of a degree of latitude near the poles. See Chapter 3 for a discussion of longitude and latitude.

TYPES OF SURVEYS

Plane Surveys

A *plane* is a flat surface, and a *plane survey* is conducted as if the earth were flat. Since the curvature of the earth has such a minimal effect over short distances, it is not a factor in plane surveys. That means that mathematical calculations in plane trigonometry and plane geometry can be used. For example, the length of an arc along the curved surface of the earth is only approximately 0.66 ft (20.12 cm) longer across the 36-mi width of a section of land than the length of a straight line, or *chord*, between the arc's endpoints. See Figure 2-2a. Since most surveys cover far smaller areas of ground, the plane survey serves well.

Another example of the offset of a plane in relation to the earth's surface is shown in Figure 2-2b. In this case, at the end of a 1-mi arc, a horizontal plane is only 0.66 ft above the earth's surface.

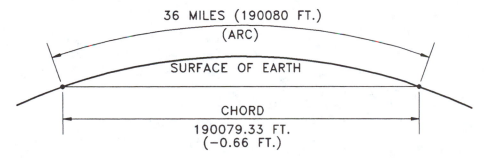

FIGURE 2–2a. The lengths of a 36-mi arc and a chord connecting the arc's ends vary by only 0.66 ft.

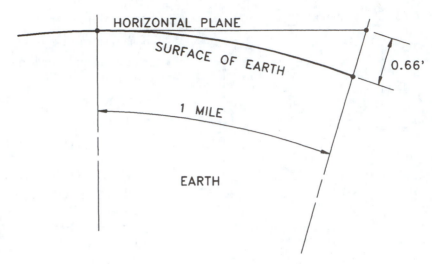

FIGURE 2–2b. A horizontal plane is only 0.66 ft above the earth's surface at the end of a 1-mi arc.

Land or Boundary Surveys

Most of us are familiar with this type of survey. If you own land, there probably exists a property plat of the plot at your county courthouse. This type of plane survey locates property corners and boundary lines. It is normally a *closed traverse* because the survey always returns to the *point of beginning* (POB) or another control point. An example of a land survey is shown in Figure 2–3.

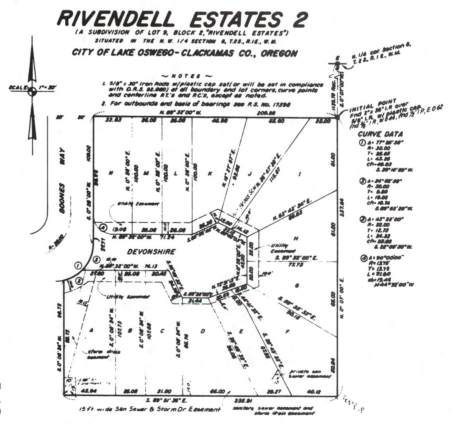

FIGURE 2–3. Typical land survey of a subdivision. (Courtesy OTAK & Associates, Inc.)

Topographic Survey

Anyone who has ever worked with a contour map has seen the results of a topographic survey. The principal function of this survey is to locate elevations and features on the land, both natural or artificial (see Figure 2–4).

Geodetic Survey

This is a grand survey, often spanning nations, in which the curvature of the earth is a factor. Large areas are mapped by a process called *triangulation*. A series of interesting triangles is established as a net. Some sides of these triangles may be hundreds of miles long, stretching from one mountain peak to another. The control established by geodetic surveys is often used as references for other surveys. Figure 2–5 illustrates the size of a typical geodetic survey.

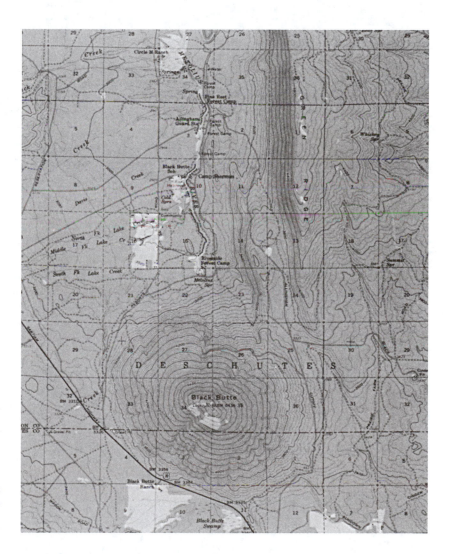

FIGURE 2–4. A topographic survey is used to compile the information needed to create this topographic map. (Courtesy U.S. Geological Survey.)

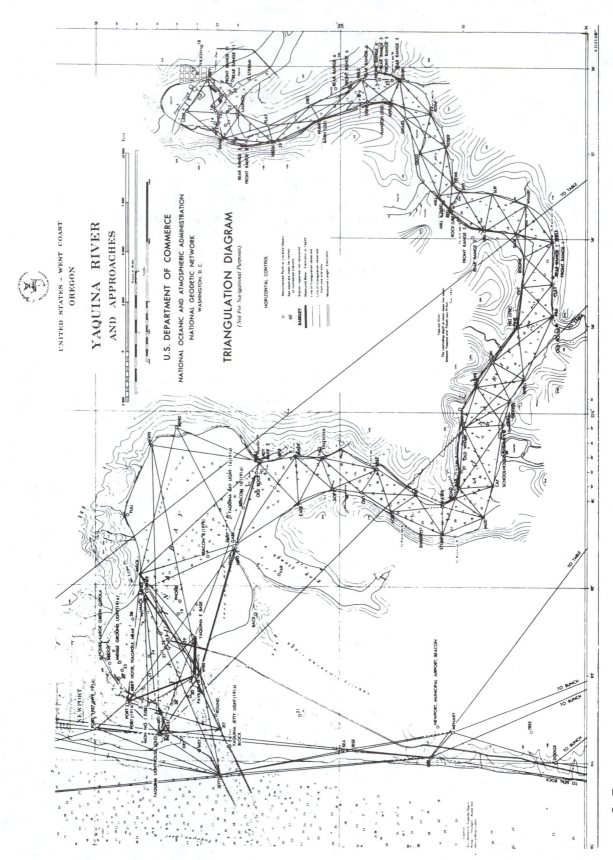

FIGURE 2–5. A geodetic survey defines major control points that can be used for smaller surveys. (Courtesy National Geodetic Survey.)

Photogrammetric Survey

Most topographic maps are now made using aerial photographs. Photographs taken at various altitudes constitute the *field notes* of this survey. Measurements are taken on the photos of known distances on the ground (often established by a land survey or open traverse) to check for accuracy, then maps are compiled using the information contained in the photograph. Many overlapping flights have to be flown before an accurate map can be created. An aerial photo used in photo mapping is shown in Figure 2–6a.

Aerial photographs are also used to compile digital maps that can then be used with a CAD system. The known distances on the ground also have known elevations. These elevations can be "seen" with a stereoscope, which utilizes our ability to detect differences in distances. With the use of a highly developed stereoscope, a photogrammetrist can not only enter data such as buildings, streets, and water edges, but also data that is dependent on elevations, such as contours, spot elevations, and depressions (see Chapter 7). The CAD operator can then access this data in digital

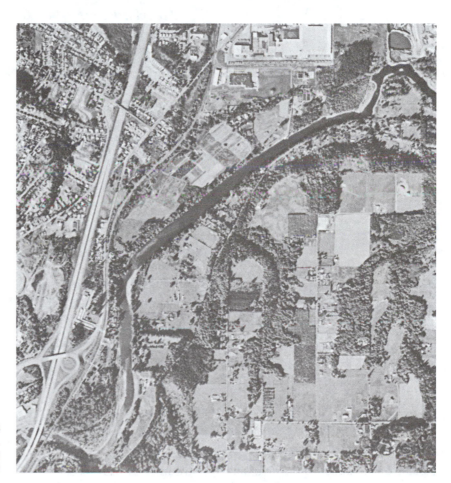

FIGURE **2–6a.** Photogrammetric surveys produce aerial photos such as this. (Courtesy Spencer B. Gross Consulting Engineer.)

FIGURE **2–6b.** A stereoscope is used to analyze stereo photos.

form by requesting information about specific lines and by reading the coordinates. A simple stereoscope is shown in Figure 2–6b.

Route Survey

An *open traverse* is a traverse that does not close on itself. An open traverse is conducted when mapping linear features such as highways, pipelines, or power lines. These are termed route surveys. They can begin at a control point such as a bench mark and consist of straight lines and angles. These surveys do not close. An example of a route survey is shown in Figure 2–7.

Construction Survey

As the name implies, the construction survey is performed at construction sites. Building lines and elevations of excavations, fills,

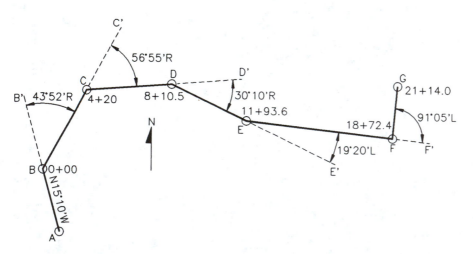

FIGURE **2–7.** Route survey or open traverse does not close on itself.

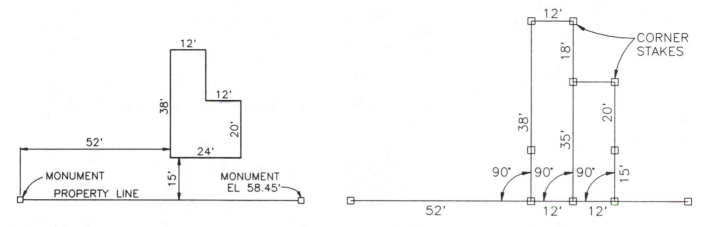

FIGURE 2–8. Construction survey showing locations of corners and staking out of house with angles and distances.

foundations, and floors are established by this localized type of survey (see Figure 2–8).

MEASURING DISTANCE

The Chain

From the rope of ancient Egypt's "rope stretchers" came the chain and the steel tape. Whereas the steel tape, seen in Figure 2–9, normally stretches to 100 ft, the old *Gunter chain* measures 66 ft. Chains of 20-m lengths, termed *land chains* because of their use in land surveys, are also popular and convenient.

FIGURE 2–9. Steel tape used to measure distance. (Courtesy Sokkia Corp.)

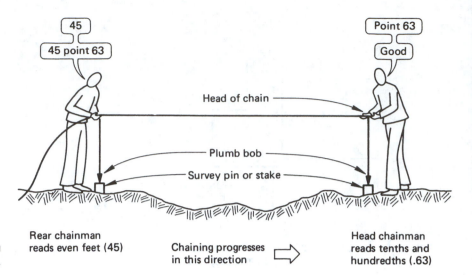

FIGURE **2–10.** Measuring distance with a chain or steel tape.

Rear chainman
reads even feet (45)

Chaining progresses
in this direction

Head chainman
reads tenths and
hundredths (.63)

The use and measurements of a chain or tape is shown in Figure 2–10. Hubs or markers of some sort are placed at each point where a reading is to be made. When chain measurements must be made on a slope, the process is often referred to as *breaking chain*, and is shown in Figure 2–11. Most chaining is now done with the steel tape and plumb bobs.

Distance by Stadia

The Greeks used the word *stadium* (plural *stadia*) when referring to a unit of length. This unit was 600 Greek feet. That translates to 606 ft 9 in. in American feet. This unit of length was used when lay-

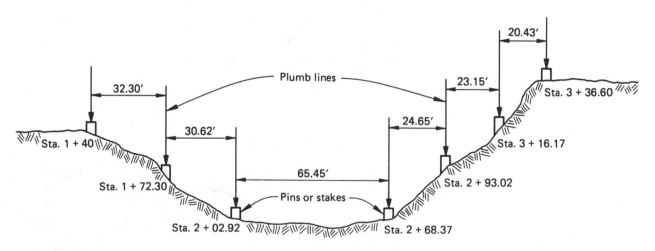

FIGURE **2–11.** "Breaking chain" on sloping terrain. Chain must be horizontal for each measurement.

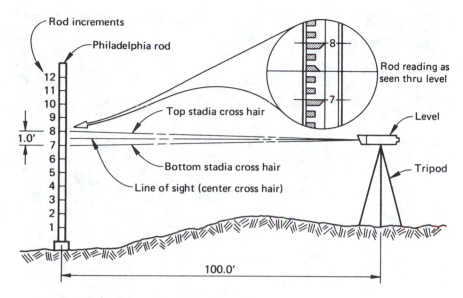

FIGURE **2–12.** Measuring distance by "stadia theory."

ing out distances in athletic contests. We now use the term to refer to a type of distance measuring employing a *rod* and an instrument with crosshairs.

The *Philadelphia rod* is 7 ft long, extends to 12 or 13 ft, and is graduated to hundreths of a foot, but can be read to thousandths. See Figure 2–19. Distances are normally read only to hundreths. Elevations can be read to thousandths. Figure 2–12 illustrates the stadia method of distance measurement, which is based on optics. The space between two horizontal crosshairs in the instrument is read by subtracting the bottom number from the top number and multiplying by 100. This gives a fairly accurate distance; one that is sufficient for low-order surveys (those requiring a lower degree of precision).

Electronic Distance Measurement

Electronic signals can be transmitted from an instrument on a tripod to a reflector on a distant tripod, which beams the signal back to the transmitting instrument. The time required for the signal to return is measured and translated into a distance. Signals such as radio waves (electromagnetic or microwave) and light waves (infrared and laser) are used in electronic distance measurement. The accuracy derived from using an electronic distance meter (EDM) is far superior, in most instances, to taping, and is much more practical in rough terrain or over long distances. Figures 2–13a and b show typical EDMs.

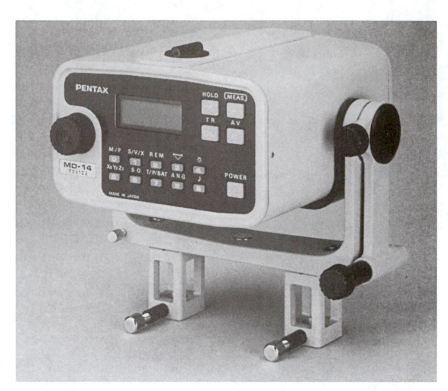

FIGURE 2–13a. An electronic distance meter (EDM). (Courtesy of Pentax.)

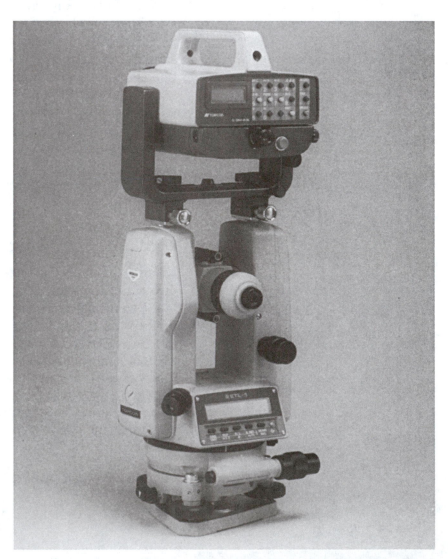

FIGURE 2–13b. An EDM mounted on an electronic theodolite. (Courtesy of Topcon.)

Elevation by Level and Rod

Electronics can take the legwork out of transferring survey data to a CAD system. Theodolites are instruments that measure horizontal and vertical angles. Many are equipped with the capability to "download" survey information gathered in the field into a computer system. The information can then be converted to language that can be understood and read by many different CAD systems. The *DXF (drawing interchange file)* format is the industry standard "language" for exchanging information between CAD systems. This DXF file can then be "imported" to most CAD systems for immediate use.

Finding the elevation of points requires only basic equipment. The process called *leveling* is normally performed using a level, a tripod, and a level rod. Figure 2–14 shows an automatic level, and Figure 2–15 shows a tripod. Simple leveling and marking of approximate elevation can be achieved using the hand level shown in Figure 2–16.

The Philadelphia rod is one of several different types of level rods, and is used with the level for measuring elevations. See Figure 2–19. The rod is placed on the known elevation and the instrument is set up approximately halfway to the unknown point. Long distances may require several setups. The tripod is anchored firmly in the ground, and the instrument is attached to the tripod. The instrument is leveled with three leveling screws on its base. A spirit bubble indicates when the instrument is level. A reading is

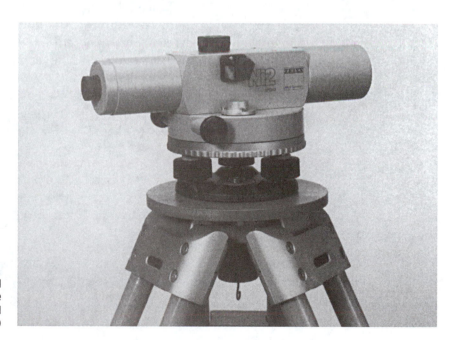

FIGURE 2–14. The automatic level maintains a horizontal line of sight once the instrument is leveled. (Courtesy of Carl Zeiss.)

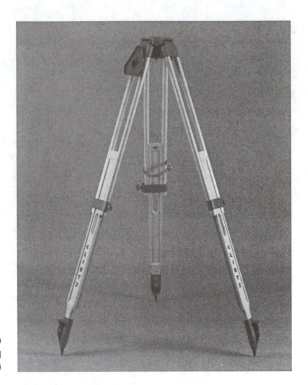

FIGURE 2–15. The tripod is used to mount levels and other surveying instruments. (Courtesy of Sokkia Corp.)

then taken on the *backsight* (rod location). This is added to the elevation of the known point to give the *height of the instrument* (H.I.). The rod is then placed on the unknown elevation (*foresight*) and a reading is taken. This reading is subtracted from the height of the instrument to give us the unknown elevation (see Figure 2–17).

Measurements that must be made over long distances require *turning points*, which are nothing more than *temporary bench marks* (TBMs) and are often as temporary as a long screwdriver driven into the ground. The turning point is just a pivot for the rod, which is used as a backsight and a foresight. The instrument reading as a foresight is subtracted from the H.I. to find the level of the TBM, and then the instrument and tripod are physi-

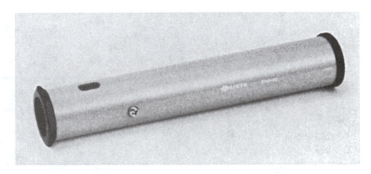

FIGURE 2–16. Hand levels are used to obtain approximate elevations. (Courtesy of Sokkia Corp.)

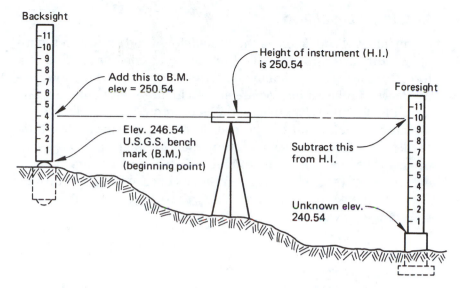

FIGURE **2–17.** Finding an elevation with level and rod (leveling).

cally moved ahead of the rod and reset. The next reading is a backsight which is added to the TBM elevation to become the H.I. Now the rod can be moved to the next foresight position. Figure 2–18 illustrates leveling with several turning points. Work your way through it.

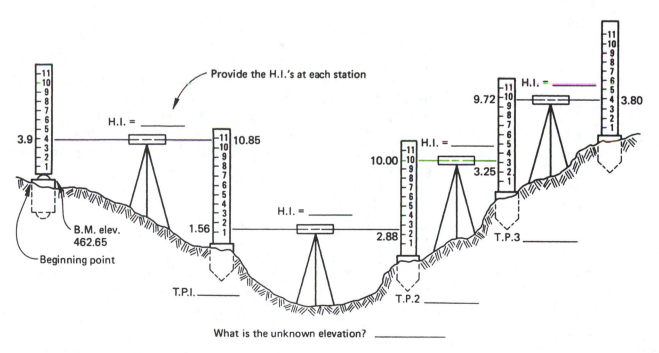

FIGURE **2–18.** Using turning points to find an unknown elevation.

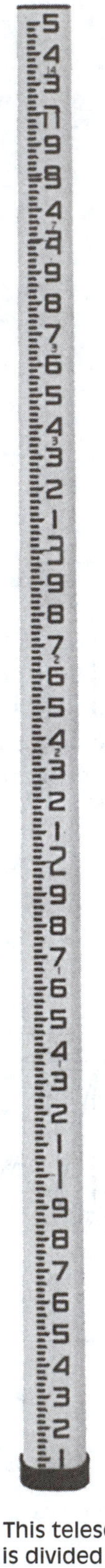

FIGURE 2–19. This telescoping Philadelphia rod is divided into whole feet and one-hundredths of a foot. (Courtesy of Topcon.)

Reading a Level Rod

A typical level rod, such as the Philadelphia rod shown in Figure 2–19, is divided into whole feet, tenths, and one-hundredths of a foot. Rods are also available in metric values. The whole foot values on the Philadelphia rod are shown in large red numerals. The large black numerals indicate tenths of a foot, and are located at a pointed black band on the rod. Each black band is one-hundredth of a foot wide. Notice the small numbers located every three tenths of a foot on the rod face. These red numbers indicate the whole foot value, and are references in case the view through the instrument shows only a small portion of the rod, and no large red number is visible.

Level Rod Targets

Rods that can be read by the person operating the instrument are called *self-reading rods*. But if elevation readings must be measured over distances longer than 200 or 300 ft, a *target* is often used. See Figure 2–20. The target is moved by the rodperson at the direction of the person running the instrument. When the target is located precisely, the rodperson can then read the graduations on the rod and target vernier and record the elevation. The vernier on the target allows measurements of 0.001 ft.

Leveling Log Books and Field Notes

Surveyors always record their measurements, instrument readings, and calculations for every project in a *field notebook*. See Figure 2–21. This notebook may become an important document for future reference, or as legal evidence. In addition, persons working in the office may have to interpret field notes in order to construct maps or drawings. Therefore field notes should be complete, neat, and uncrowded. The following items are important in field notes:

- Title of the survey
- Members of the survey crew
- Location
- Frequent sketches
- Instrument numbers
- Date of survey
- Duty of each crew member
- Weather
- Mistakes crossed out, not erased

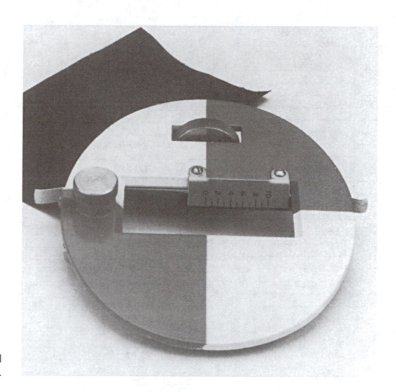

FIGURE **2–20.** The vernier on the level
rod target reads to 0.001 ft.

The type of survey determines the content of the field notes. For example, field notes for a traverse may contain deflection angles, bearings, distance on slope, vertical angles, and horizontal distances. This information can then be used by a computer program to generate a database and a drawing, or it can be used in the

FIGURE **2–21.** Field notebooks used by
surveyors. (Courtesy of Pentax.)

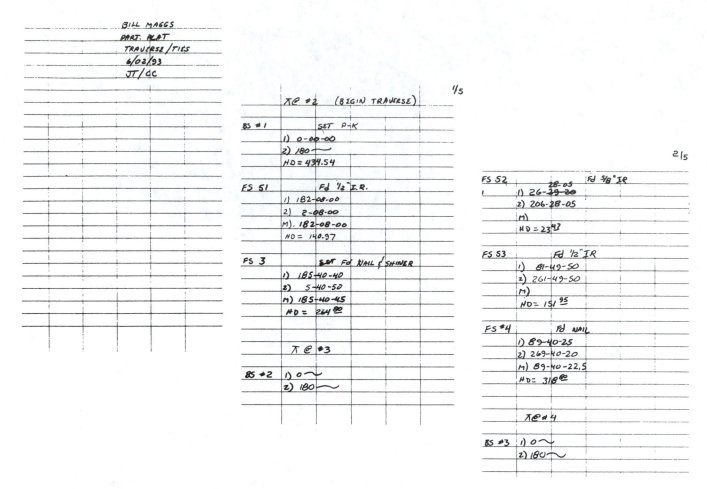

FIGURE 2–22a. These field notes of a closed property traverse indicate that several shots were taken from each instrument setup. The 1) is the initial angle reading; 2) is the check angle looking back from the foresight; M) is the average of the two, or "mean"; and HD is the distance from the instrument to the point being measured. (Courtesy of Centerline Concepts, Inc.)

manual construction of a drawing. Figure 2–22a shows an example of this type of field notes.

Field notes based on leveling contain only location and elevation information. Look at the field notes in Figure 2–22b. The left column contains the location of the rod. The B.M. indicates a benchmark, or permanent marker. The T.P. represents a turning point. This is a temporary point on which the rod is rested in order for the instrument to get a reading. It is called a turning point because, once an elevation has been determined for the T.P., the instrument is physically moved to the next station point. The rod is then physically turned to face the instrument. The B.S. is the backsight, and is the reading taken looking back toward the point of beginning. The H.I. is the height of instrument, and is calculated

FS #54		Fd 1/2" IP, BENT, TIED BLND
	1) 29-29-20	
	2) 209-29-20	
	M)	
	HD= 34 03	
FS 55		Fd 5/8" IR, SMOOTH + SLOTTED
	1) 76-51-50	
	2) 256-52-00	
	M) 76-51-55	
	HD= 128 20	
FS #5		SET P+K
	1) 85-27-05	
	2) 265-27-20	
	M)	
	HD= 262 03	
	π@ #5	
BS #4	1) 0 ~	
	2) 180 ~	

FS #6		SET P+K IN S/W CRACK
	1) 182-08-20	
	2) 2-08-40	
	M) 182-08-30	
	HD = 366 64	
	π@ #6	
BS #5	1) 0 ~	
	2) 180 ~	
FS S6		
	1) 91-08-20	
	2) 271-08-20	
	M) 91-08-20	
	HD= 139 73	
FS #7 =#1		
	1) 106-19-00	
	2) 286-19-00	
	M) 106-19-00	
	HD = 332 15	

	π@ #7 = #1	
BS #6	1) 0 ~	
	2) ~ 180 ~	
FS #2	1) 70-44-20	
	2) 250-44-20	
	M) 70-44-20	
	CLOSING ANGLE	
	END TRAVERSE	

FIGURE 2–22a. (continued)

by adding the reading taken from the backsight. After the backsight reading is taken, the rod is physically moved forward and set up on the next T.P. The next reading, called the foresight (F.S.), is taken and subtracted from the H.I. to give the elevation of the new T.P., which is listed in the last column. The leveling process continues until the final elevation is determined.

The difference between the totals of the backsights and the totals of the foresights equals the difference in elevation between the bench mark and the final elevation. This value can be used to check for error in the intermediate calculations.

Construction Leveling and Grading by Laser

Construction work usually requires the use of leveling to set elevation grade stakes for road, parking lot, and driveway grades, as well as elevations for excavations, fills, concrete work, plumbing,

Thomas & Nancy Doherty
13605 SE Clack. Riv. Dr.
FLOOD ELEV.
12/10/92
JT/JL
OVERCAST, RAIN, WINDY
± 50°F

POINT	ZENITH ∆ DIR	ZENITH ∆ REV	SLOPE DIST	∆ ELEV	ELEV	½
BS B.M. E635	88-57-48	271-02-05	1096.35	-19.817	76.515	
FS TP #1	89-33-45	270-26-00	802.63	+6.092		
				-13.725		
BS TP #1	89-31-28	270-28-20	603.99	-5.012		
FS TP #2	89-16-40	270-43-07	342.68	+4.309		
				-0.703		
BS TP #2	89-25-27	270-34-18	1382.40	-13.843		
FS TBM	92-27-07	267-32-32	130.99	-5.611	42.612	
SW COR CONC SLAB @ BASE STAIRS – RIVER SIDE OF BSL.				-19.454		
FS GAR. SLAB (LOW POINT)	92-54-21	267-05-29	112.53	-5.707	42.516	
FS FF (MAIN FLOOR)	86-51-44	273-07-57	161.88	+8.853	57.076	
	RUN		CHECK			
	-13.725		+19.865			
	-0.703		+0.429 / +14.630			
	-19.454		-1.000	AVG ∆ELEV = 33.903		
	-33.882		+33.924			
			33.882			
			0.042			

FIGURE 2–22b. These field notes are of differential leveling and angle measurements done with an EDM. These notes consist of a zenith angle, both direct and reverse, distance, and elevation.

and floor level verification. The electronic, or laser, level is used for these purposes. The *rotating beam* laser is commonly used because it projects a plane of reference as the beam rotates 360°. The laser beam can be picked up on special *beam detectors* fitted to rods or telescoping detector poles. Figure 2–23 shows an electronic level and beam detector.

The laser can be set either to provide a level plane or to a single beam for use in pipelines and tunnels. The beam or plane can also be tilted to provide a slope for establishing parking lot and road grades. Laser beams can be effective to a radius of more than 600 ft.

FIGURE 2–23. The laser level on the left projects a rotating beam as a plane of reference. The beam detector on the right mounts to a rod to measure elevation. (Courtesy of Sokkia.)

TRAVERSING

A *traverse* is a series of continuous lines connecting points called *traverse stations* or *station points*. The lengths of the lines connecting the points are measured, as are the angles between the lines. Several traverse types are currently in use.

Open Traverse

The open traverse, as seen in Figure 2–24 and mentioned previously, consists of a series of lines that do not return to a POB and do not necessarily have to begin or end on a control point. Exploratory surveys employ this type of traverse where accuracy is not critical, and estimates will satisfy the project requirements. The open traverse cannot be checked easily and is not suitable for work other than route surveys.

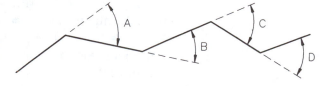

FIGURE 2–24. The open traverse does not begin or end at a control point and cannot be easily checked.

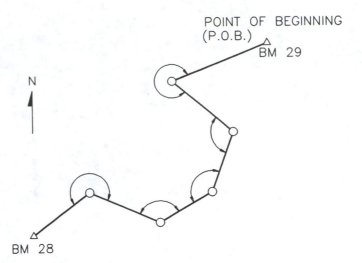

POINT OF BEGINNING
(P.O.B.)
△
BM 29

N

△
BM 28

FIGURE 2–25. A connecting traverse is one in which the beginning and ending points are known.

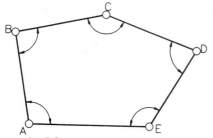

FIGURE 2–26. A loop traverse closes on itself and is easily checked.

Connecting Traverse

Another form of traverse is the *connecting traverse* in which the beginning and ending points are known. This type of traverse can result in an accurate survey because both angular and linear measurements can be checked to determine location. See Figure 2–25. This appears to be an open traverse, but is technically a closed traverse because it connects known points and is easy to check.

Closed Traverse

In the closed type of traverse, the lines close on the point of beginning, as in a *loop traverse*, or close on a different known control point, as in the *connecting traverse*. The closed traverse can be checked for accuracy and is thus used exclusively for land surveys and construction surveys. Figure 2–26 shows an example of a loop traverse.

Compass Bearing Traverse

Bearings are angular measurements of 0° to 90° taken from a north or south line and are oriented either east or west. The surveyor's compass was originally used when laying out a traverse, and the bearings were read directly from the compass. In a present-day compass bearing traverse, bearings are calculated using such instruments as the transit, theodolite, and total station. (See Figures 2–27a–e.) The electronic total station is a complete instrument that measures horizontal and vertical angles, and contains a built-in EDM. Data recorded in an electronic total station can be saved in an electronic field book or field computer (right side,

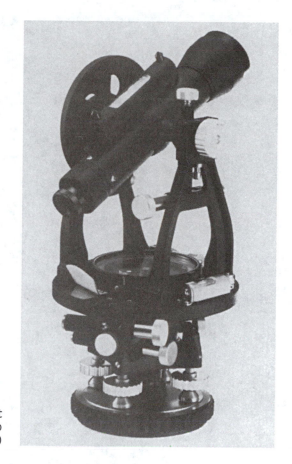

FIGURE 2–27a. The one-minute transit measures horizontal and vertical angles to one minute accuracy. (Courtesy of Sokkia.)

Figure 2–27d). Data can then be downloaded at the office into a computer or transmitted directly from the site to the office. Bearings measured with these instruments are more accurate, and can be used with great precision in calculations for mapping. The bearing of the backsight is known and the angle is then measured to the foresight. This angle is applied to the back bearing to determine the foresight bearing. This method is often used in connecting traverses, as shown in Figure 2–25.

Direct or Interior Angle Traverse

This is the principal method by which closed traverses are measured. The traverse proceeds either clockwise or counterclockwise around the plot and measures the interior angles. The closed traverse shown in Figure 2–26 is plotted by the direct angle method. Bearings can later be applied to the direct angles if required.

An interior angle traverse can be run clockwise or counterclockwise. Regardless of the direction of the survey, the sum of

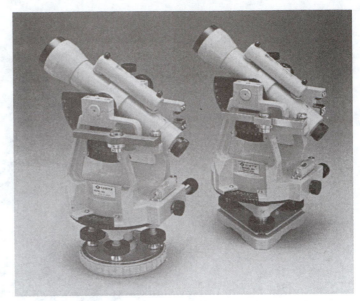

FIGURE 2–27b. The engineering and construction transit measures horizontal and vertical angles to 20" accuracy. (Courtesy of Sokkia.)

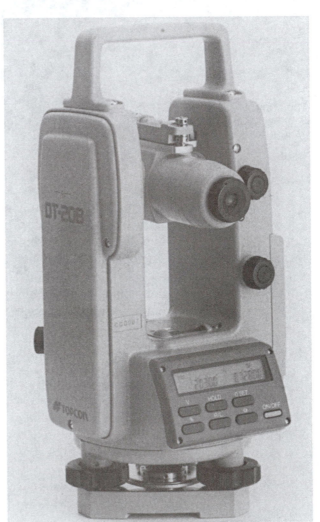

FIGURE 2–27c. The electronic digital theodolite measures horizontal and vertical angles to 20" accuracy. (Courtesy of Topcon.)

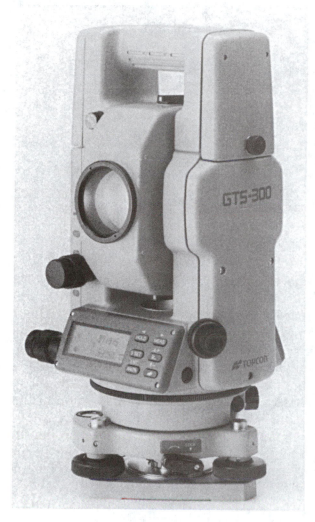

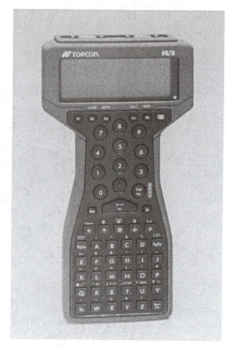

FIGURE **2–27d.** The electronic total station combines vertical and horizontal angle measurements with an EDM, and can be cabled to a handheld field computer. (Courtesy of Topcon.)

the interior angles should always equal $(n - 2)180$, where n is the number of sides in the polygon.

Deflection Angle Traverse

A deflection angle is one that veers to the right or left of a straight line (see Figure 2–28). This method of angle measurement is commonly used in route surveys. An angle to the right or left of the backsight is measured and is always 180° or less. The letters R or L must always be given with the angle.

When the preceding line of a survey is extended beyond the current station point, it is said to be a *prolongation* of that line.

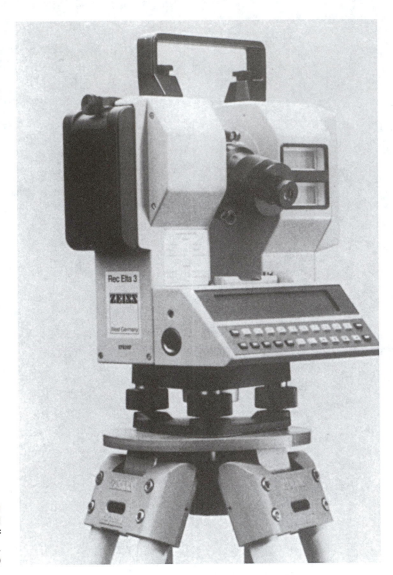

FIGURE 2–27e. This total station measures temperature and pressure, and uses this data for automatic correction of distance and elevation measurements. (Courtesy of Carl Zeiss.)

The deflection angle is measured from the prolongation of the succeeding line. But the surveyor does not just measure a 180° horizontal angle from the preceding line, because improper instrument adjustment may lead to error. Instead, the instrument is aimed at the backsight and then flipped on its transverse, or horizontal axis, to achieve the prolongation. Then the deflection angle is measured to the foresight (next station point). When an instrument is flipped, or turned over on its transverse axis, it is said to be *plunged*. A deflection angle will close when the difference between the angles to the right and the angles to the left equals 360 degrees.

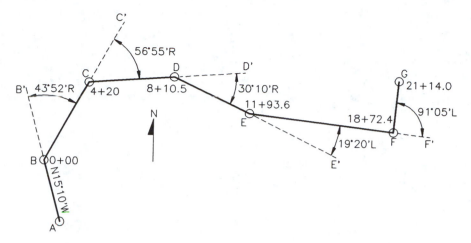

FIGURE 2–28. Deflection angle traverse.

Angles to the Right

This type of survey always measures the foresight by turning the instrument clockwise from the backsight. When the traverse route is clockwise, the angles to the right are exterior angles, or outside the polygon. The sum of these angles should total $(n + 2)180°$. A traverse route that is counterclockwise creates interior angles inside the polygon. The sum of these angles should total $(n - 2)180°$. Figure 2–29 shows the appearance of these two types of angles to the right traverses.

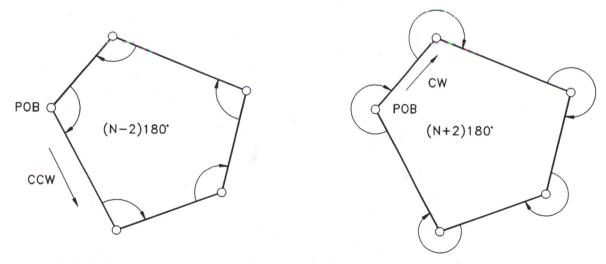

FIGURE 2–29. The foresight is measured by turning the instrument clockwise from the backsight in an angle to the right survey.

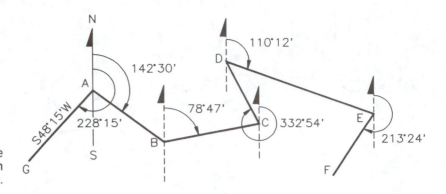

FIGURE 2–30. An azimuth traverse measures each angle clockwise from north or south.

Azimuth Traverse

Azimuth means horizontal direction. It is derived from the Arabian word "al-samt," which means "the way." In mapping, we use the term to refer to a direction that is measured from a north or south line. Unlike bearings, which measure only 90° quadrants, the azimuth is a measurement that encompasses the entire 360° of a circle. An azimuth traverse requires only one reference line; that is most often a north-south line. This reference line can be either true or magnetic. In Figure 2–30, the azimuths are measured clockwise from the north line.

ELECTRONIC TRAVERSING

Measuring distances and elevations, as well as plotting traverses, are not usually accomplished in an isolated setting. In other words, surveys are begun by initially tying them either to previous surveys or to an organized reference grid or coordinate system. Many survey points have been established across the globe, allowing for a precise network of measurements. Grids of this kind enable us to accurately reflect curved land masses.

In the United States, two grid systems are used: the Lambert Conformal Projection and the Transverse Mercator Projection. These two systems are known as State Plane Coordinate (SPC) Systems. An example of grid systems is shown on the edge of the Mount Hood North Quadrangle map in Figure 2–31.

Surveys of this nature are important in CAD work. With both aerial topographic maps and surveys being produced with the same SPC reference, the CAD operator is able to overlay the aerials and surveys with a high degree of accuracy, using the SPC coordinates for both types of maps. Refer to Chapter 3 for a more in-depth look at grid systems.

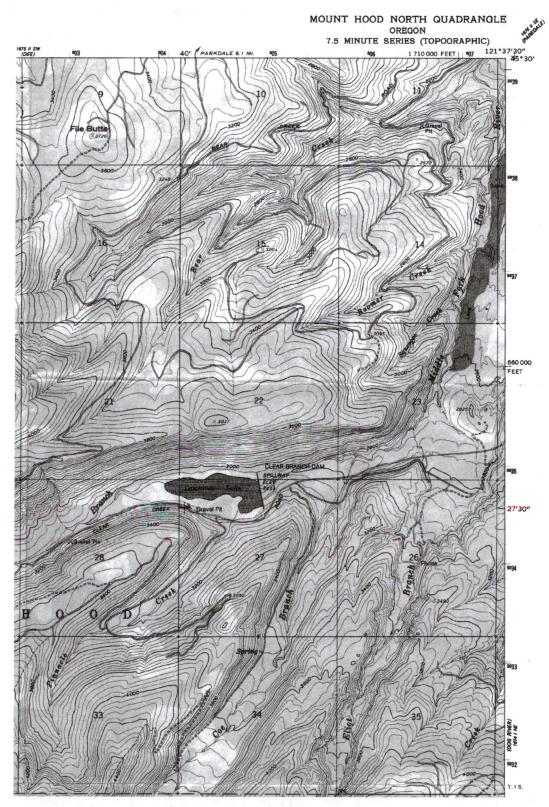

FIGURE 2–31. A topographic map showing State Plane coordinates and UTM rectangular coordinates. (Courtesy of United States Geological Survey.)

GLOBAL POSITIONING SYSTEM

The Pentagon has established an orbital network of 21 geosynchronous satellites, each of which transmits radio signals to earth. The network is constructed so that four satellites are in view from any position on the earth. A surveyor on earth, using a Global Positioning System (GPS) receiver (Figure 2–32) can determine longitude and latitude using the signals from three satellites. The elevation of a point can be determined using a fourth satellite.

The military degrades the signal used for civil purposes to achieve the Standard Positioning Service (SPS). This provides accuracy of approximately 100 m. But these inaccuracies can be calculated by using a technique called *differential GPS*. Using this system, GPS receivers are placed on known points, such as bench marks. Then the GPS positions are measured and the differences between the known points and the GPS values are used to calculate the amount of error (differential) that can be used in subsequent GPS measurements.

The uses of GPS extend not only to surveying, but to a wide range of civil and private applications. It can be used in motor vehicles, ships, aircraft, and pleasure boats for locating position. It can also be used for accurate measurement of elevations, as well as volcanic and seismic activity. As the system is improved and maintained, it may become a vital part of our overall measurement system, and its applications may be endless.

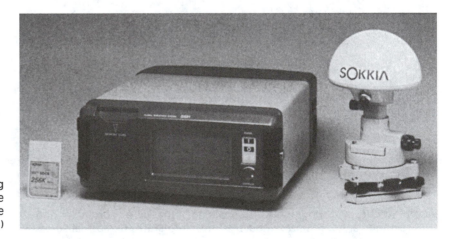

FIGURE 2–32. A global positioning surveying system is used to determine longitude and latitude from satellite transmissions. (Courtesy of Sokkia.)

TEST

2–1   What is a topographic survey?_____

_____

2–2   In What type of survey is triangulation used? _____

_____

2–3   A route survey is often termed an open traverse. Why? _____

_____

2–4   What are two things the Philadelphia rod can be used for? _____

_____

2–5   What is "breaking chain"? _____

_____

2–6   Would a turning point be used to locate (a) elevation or (b) distance?

_____

_____

2–7   What type of survey would be used to lay out a new highway? _____

_____

2–8   Can the type of survey in question 7 be checked easily? _____

_____

2–9   What is a station point? _____

_____

2–10   What type of instrument is used to measure bearings? _____

_____

2–11   An angle to the right or left of the backsight is termed a _____

_____

2–12   Compare and contrast bearings and azimuths _____

_____

2-13 What is the difference in miles between the equatorial and polar diameters of the earth? _____

_____

2-14 What type of mathematical calculations are used in a plane survey?

_____

_____

2-15 What is EDM? _____

_____

2-16 What is the smallest increment marking on a Philadelphia rod?

_____

_____

2-17 What do the small red numbers on a Philadelphia rod represent?

_____

_____

2-18 How accurate can level rod measurements be when a rod target is used? _____

_____

2-19 What type of instrument is used for construction grading? _____

_____

2-20 What is the formula for calculating the sum of the angles of an interior angle traverse? _____

_____

2-21 What is the system called that uses satellites to locate points on the earth's surface? _____

_____

2-22 Identify the following types of traverses:

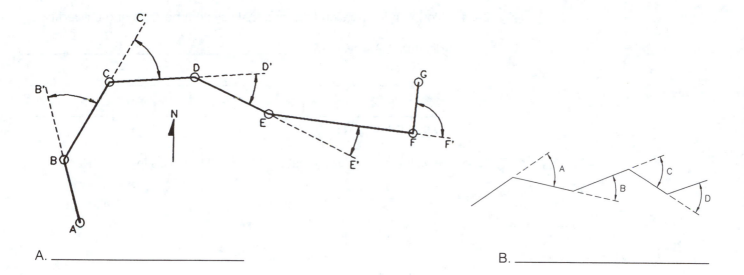

A. _____

B. _____

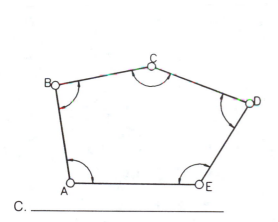

C. _____

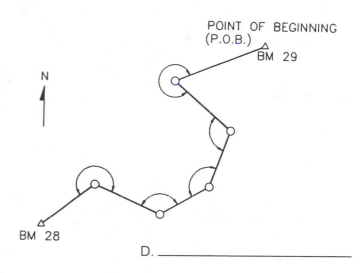

POINT OF BEGINNING
(P.O.B.)
△ BM 29

N

BM 28 △

D. _____

2–23   What type of map is shown in Figure 2–31?_____

_____

2–24 How can information from one CAD system be read and utilized by a different CAD system? _____

_____

PROBLEMS

P2–1 Give the elevations indicated by the rod readings in Figure P2-1.

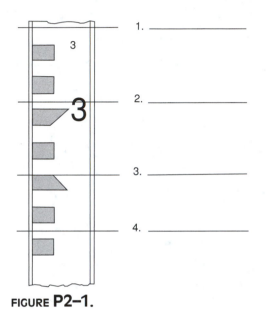

1. _____

2. _____

3. _____

4. _____

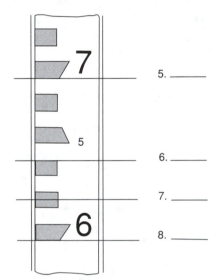

5. _____

6. _____

7. _____

8. _____

FIGURE **P2–1.**

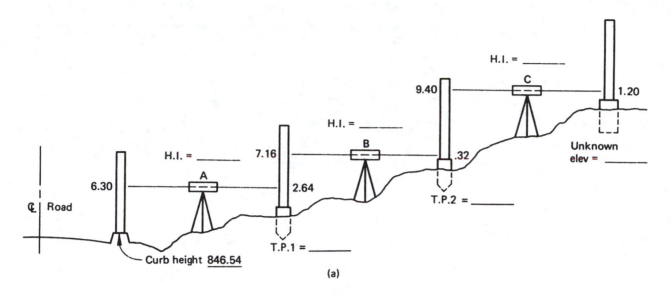

(a)

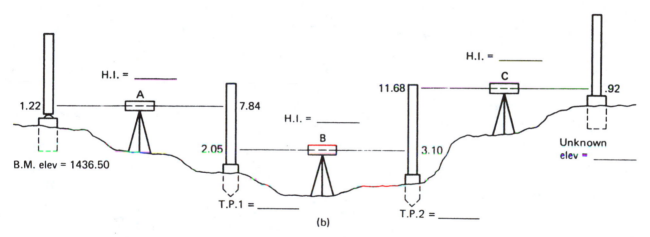

(b)

FIGURE **P2–2.**

P2–3 Calculate the unknown elevations given the field notes provided in Figure P2-3.

(Note that the foresight to TP-1 from the first instrument setup is shown on the second line of the notes, on the same line as the label TP-1.)

Station	Backsight (+)	H.I.	Foresight (−)	Elevation
US.GS.B.M.I	6.21			422.34
T.P.1	8.90		3.46	
T.P.2	7.82		4.63	
T.P.3	5.60		2.45	
B.M.2	12.04		3.86	
T.P.4	4.62		7.70	
T.P.5	7.31		4.06	
B.M.3			3.64	

FIGURE **P2–3.**

Problems 2–4 to 2–6 give you incomplete level field notes. Complete the field notes, adding all missing H.I.s and elevations. Check your final answer mathematically, as discussed in the chapter.

P2–4

Station	BS(+)	HI	FS(−)	Elevation
BM-1	3.63			247.4
TP-1	6.24		5.29	
TP-2	8.65		6.04	
TP-3	2.51		10.16	
BM-2			7.55	

P2–5

Station	BS(+)	HI	FS(−)	Elevation
BM-1	8.21			946.85
TP-1	5.36		4.89	
TP-2	7.81		10.63	
TP-3	9.07		6.32	
TP-4	5.11		4.99	
BM-2			7.25	

P2–6

Station	BS(+)	HI	FS(−)	Elevation
BM-1	6.992			946.85
TP-1	7.964		9.376	
TP-2	3.560		4.096	
TP-3	7.455		3.675	
TP-4	3.442		7.521	
TP-5	8.459		12.680	
TP-6	9.456		3.442	
TP-7	6.087		4.358	
BM-2			2.961	

Problems 2–7 and 2–8 give you illustrations of leveling. Using the information given, set up appropriate field notes and provide all necessary information. Solve the problems for all H.I.s, elevations, and final BM-2 elevations. Provide a mathematical check of your work in the field notes.

P2–7

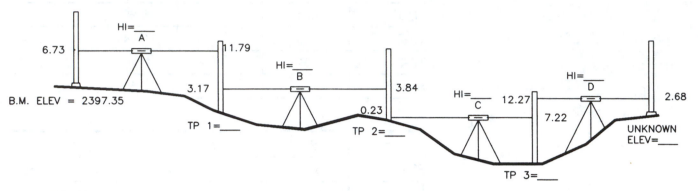

FIGURE P2–7.

P2–8

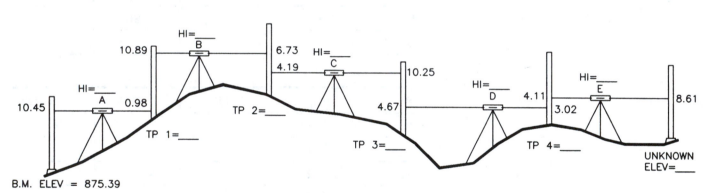

B.M. ELEV = 875.39

FIGURE P2–8.

In Problems 2–9 and 2–10 you are given illustrations of plan views of two differential leveling operations. Prepare appropriate field notes including all elevations and a math check for each problem. The small circles along the level line indicate instrument setups, and the numbers are foresight and backsight readings from the instrument.

P2–9

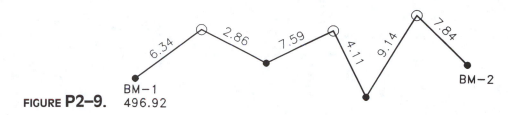

FIGURE **P2–9.**

BM–1
496.92

P2–10

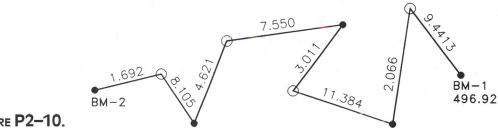

FIGURE **P2–10.**

Location and Direction

Location and direction are two of the main purposes of map-making and use. This chapter explains the division of the earth into parts and how this system can be used to locate features. Also discussed are basic map geometry and construction of plats.

The topics covered include:

- Longitude
- Latitude
- Location on a map
- Direction
- Azimuth
- Bearing
- Map geometry
- Traverse

LOCATION

The earth has been gridded by imaginary lines, called lines of *longitude* and *latitude*. These lines were established to aid location of features on a map.

If the earth were rectangular in shape, we could use a rectangular system that would be easy to measure and each square would be the same. With the earth spherical as it is, we have established points from which measurements can be accurately made. These points are the north and south poles, and the center of the earth. From these points, a grid system has been established using the degrees of a circle as reference. The grid lines are referred to as lines of longitude and latitude.

Longitude

Lines of longitude are imaginary lines that connect the north and south poles. These lines are also referred to as *meridians*. The imaginary line connecting the north and south poles, and passing through Greenwich, England, is called the *prime meridian*. The prime meridian represents 0° longitude, as shown in Figure 3–1. There are 180° west and 180° east of the prime meridian, forming the full circle of 360°. Meridians east of the prime meridian are referred to as *east longitude*. Meridians west of the prime meridian are called *west longitude*. The 180° meridian coincides roughly with the *International Date Line*.

Length of a Degree of Longitude.

A degree of longitude varies in length at different parallels of latitude, becoming shorter as the parallels approach the north and south poles. That is a little confusing, but think about it for a bit and refer back to Figure 3–1. There you can see how degrees of longitude get closer together. Table 3–1 shows the length of a degree of longitude at certain latitudes.

TABLE 3–1. Length of a degree of longitude

Latitude	Statute Miles	Latitude	Statute Miles
0°	69.186	50°	44.552
10°	69.128	60°	34.674
20°	65.025	70°	23.729
30°	59.956	80°	12.051
40°	53.063	90°	0.000

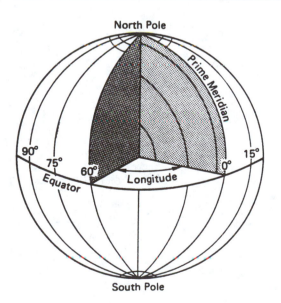

FIGURE 3–1. Measuring longitude.

The length of a degree of longitude may be calculated using this formula:

Length of a degree of longitude =
  Length of a degree of latitude × Cosine latitude

The length of a degree of latitude is approximately 69 statute miles. Calculate the length of a degree of longitude at, for instance, 45° latitude in the following manner:

Length of a degree of longitude = 69 statute miles × cosine 45°
                                 = 69 × 0.70711
                                 = 48.791 statute miles

Latitude

Latitude is measured as an angular distance from the point at the center of the earth. Look at Figure 3–2 for an illustration. Notice that the latitude is the angle between the line of the equator and the other side of the angle.

Points on the earth's surface that have the same latitude lie on an imaginary circle called a *parallel of latitude*. Lines of latitude are identified by degrees north or south of the equator. The equator is 0° latitude. The north pole is 90° *north latitude* and the south pole is 90° *south latitude*.

Length of a Degree of Latitude. Parallels of latitude are constructed approximately the same distance apart. Due to the bulging of the earth near the equator, each degree of latitude is not absolutely the same in distance. A degree of latitude varies from 69.4 statute miles near the poles to 68.7 statute miles at the equa-

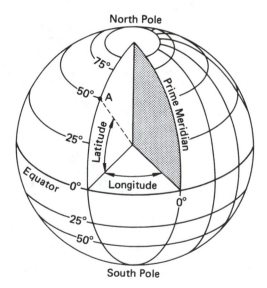

FIGURE **3–2.** Measuring latitude.

tor. For all but the very detailed maps, it is satisfactory to refer to each degree of latitude as being 69 statute miles in length. A *statute mile* is established as an international standard and intended as a permanent rule. A statute mile has 5,280 feet.

Location on Maps

On flat maps, meridians and parallels may appear as either curved or straight lines. Most maps are drawn so that north is at the top, south at the bottom, west at the left, and east at the right. Look for the north arrow for the exact orientation.

To determine directions or locations on any particular map, you must use the coordinates of parallels and meridians. So if you want to know which city is farther north, Portland or Salem, the map in Figure 3–3 shows you that Portland is north of Salem. Which city is farther east? Portland or Salem? The map tells the story. Salem is on the 123° meridian and Portland is east of that. The point that is actually being made is that any location on the earth can be identified by locating its intersecting lines of longitude or latitude.

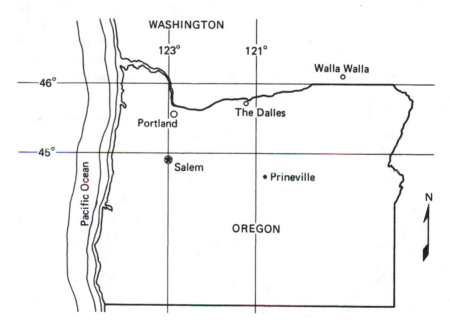

FIGURE 3–3. Using coordinates of parallels and meridians to find locations.

DIRECTION

A *direction*, in surveying, refers to the angular relationship of one line to another. When a number of lines radiate from a point, the direction of these lines is expressed with reference to one of the lines that is designated as having zero direction. In most cases a north-south or east-west line carries the zero designation.

Units of Angular Measure

Units of angular measure are *degrees*. Degrees are identified with the symbol °, as in 30°. There are 360° in a complete circle, hence a quarter circle has 90°, and so on. Each degree is made up of 60 minutes. The minutes symbol is ′. Minutes are divided into 60 seconds, which are identified as ″. So a complete degree, minute, and second designation may read like this: 50°30′45″. It is important when you are doing calculations with degrees and parts of degrees that you keep this information in mind. Consider these examples:

$$
\begin{array}{r}
48°40′25″ \\
+\ 25°38′40″ \\
\hline
73°78′65″
\end{array}
\qquad
\begin{array}{r}
\text{borrow 1}° \qquad\quad {}^{4\ 9\ 2} \\
7\cancel{5}°\cancel{32}′10″ \\
-\ 34°45′\ 8″ \\
\hline
40°47′\ 2″
\end{array}
$$

reduced to 74°19′5″.

Surveyor's Compass

The surveyor's compass is used in mapping to calculate the direction of a line. The reading taken is usually a bearing angle or an included angle. Figure 3–4 shows the differences between two commonly used compasses: the mariner's compass and the surveyor's compass.

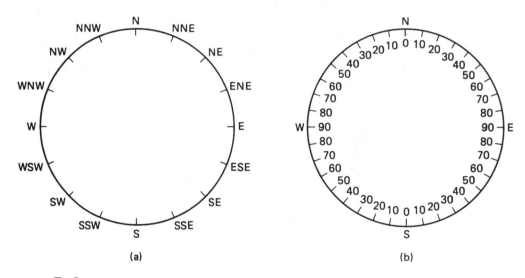

(a)   (b)

FIGURE 3–4. (a) Mariner's and (b) surveyor's compass.

Azimuth

An *azimuth* is a direction, measured as a horizontal angle from a zero line, generally north-south, in a clockwise direction (see Figure 3–5).

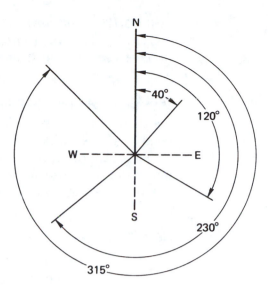

FIGURE **3–5.** Azimuth.

True Azimuth

A *true azimuth* is a horizontal angle measured using true north as the reference line.

Grid Azimuth. A *grid azimuth* is established for a rectangular survey system so that the north-south grids of the survey are used as the reference, or zero line. You can see this used when you work with the township section system. Refer to the explanation about UTM and SPC grids later in this chapter.

True North (Geographic North). Abbreviated TN, *true north* is the location of the North Pole. *Geographic north, geographic meridian,* and *true meridian* are terms that are all used to designate the same meaning as true north.

Magnetic North. *Magnetic north (MN)* is where the compass north arrow points. Magnetic north is about 1000 miles away from true north. Sometimes the terms *magnetic compass* and *magnetic meridian* are used to mean magnetic north.

Grid System. *Grid systems* are used to establish points of reference for features of the earth's surface when preparing map drawings. Coordinate systems have been developed to help ensure that the curvature of the earth coincides with the rectangular grid on a map drawing. The most commonly used grid systems are the Universal Transverse Mercator (UTM), the Universal Polar Stereographic (UPS), and the State Plane Coordinate (SPC). When you see a note on a drawing, such as UTM grid and 1994 magnetic north, this means that the grid north given refers to the Universal Transverse Mercator grid system.

Grid North. *Grid north (GN)* consists of the north-south lines of a grid mapping system previously discussed. The angular variation between true north and grid north is often given in a map legend.

Magnetic Azimuth. *Magnetic azimuths* are measured with magnetic north as the zero line. Actually, the magnetic compass is used only as a check on more accurate methods and as a method to obtain approximate values for angles. The magnetic azimuths may differ from the true azimuth by several degrees, depending on the local magnetic attractions.

Bearings

The *bearing* of a line is its direction with respect to one of the quadrants of the compass. Bearings are measured clockwise or counterclockwise, depending on the quadrant, and starting from north or south (see Figure 3–6).

A bearing is named by identifying the meridian, north or south; the angle; and the direction from the quadrant, east or west. Therefore, a line in the northeast quadrant with an angle of 30° with the north meridian will have a bearing of N30°E. Consider a line in the northwest quadrant that is 40° from north; the bearing will be N40°W. Look at the other examples of bearings and azimuths in Figure 3–7.

Magnetic Declination

The meridian indicated by the needle of a magnetic compass seldom coincides with the true meridian. So the horizontal angle between the magnetic meridian and the true meridian at any point is called the *magnetic declination*. The magnetic declination is either east or west, depending on the direction the arrow of the

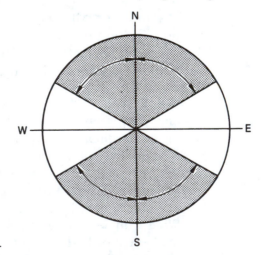

FIGURE 3–6. Bearing.

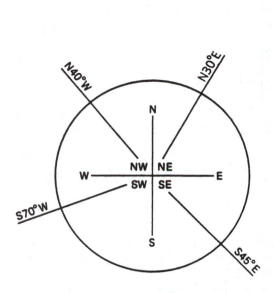

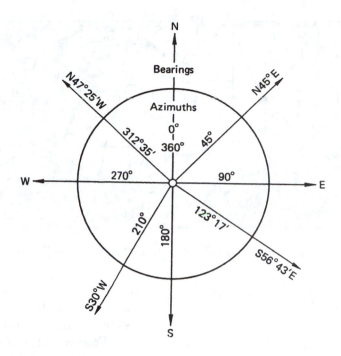

FIGURE **3–7.** Examples of bearings and azimuths.

compass points from true north. The magnetic declination of a map must be updated periodically because of continuous changes in its value. Figure 3–8 shows an example of a compass reading displaying the angle between geographic north and magnetic north as the magnetic declination. Magnetic declination is measured in degrees and mils. *Mils* are units of angular measure where one mil equals 1/6400 of the circumference of a circle. Figure 3–9 displays an illustration of grid north and magnetic north in relationship to true north. This example was taken from

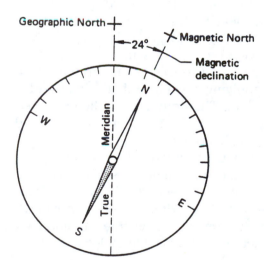

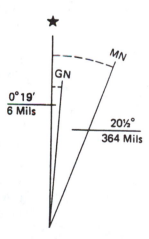

FIGURE **3–8.** Sample of magnetic declination.

FIGURE **3–9.** UTM grid and 1994 magnetic north declination at center of sheet.

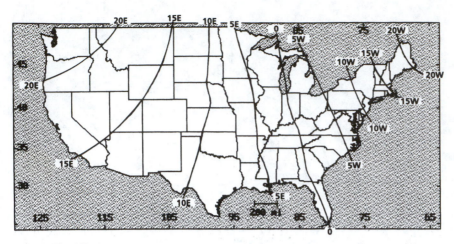

FIGURE 3–10. Magnetic declination changes throughout the United States.

the legend of an actual map. Figure 3–10 shows a good representation of how the magnetic declination changes throughout different locations in the United States.

Following are some abbreviations and terms associated with making magnetic declination calculations:

MD—magnetic declination

EMD—east magnetic declination

WMD—west magnetic declination

EMA—east magnetic azimuth

WMA—west magnetic azimuth

TA—true azimuth

Figure 3–11 demonstrates how to calculate the true azimuth, given several different values of magnetic declination.

When a drafter draws bearing lines from the surveyor's notes, they are either true bearings or magnetic bearings. If they are true bearings, the surveyor notes this in the field book: "Bearings are referred to the true (north-south) meridian." The bearings are corrected by applying the magnetic declination for the correct time (year and day). The drafter plots the corrected bearings as given in the field notes. The completed map should indicate if the bearings are true bearings.

Most maps carry the magnetic declination for a specific year, as well as the amount of annual change in degrees. The change is indicated as easterly or westerly, and records whether the compass is pointing east of true north (easterly), or west of true north (westerly). After calculating the annual change, the amount must be either added to or subtracted from the angle of declination.

Local Attraction

A *local attraction* is any local influence that causes the magnetic needle to deflect away from the magnetic meridian.

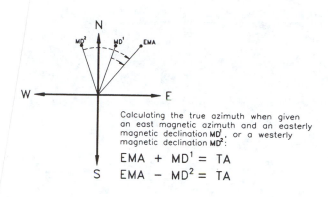

Calculating the true azimuth when given
an east magnetic azimuth and an easterly
magnetic declination MD¹, or a westerly
magnetic declination MD²:

$$EMA + MD^1 = TA$$
$$EMA - MD^2 = TA$$

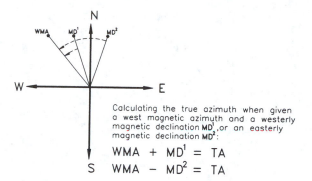

Calculating the true azimuth when given
a west magnetic azimuth and a westerly
magnetic declination MD¹,or an easterly
magnetic declination MD²:

$$WMA + MD^1 = TA$$
$$WMA - MD^2 = TA$$

FIGURE 3–11.

Local attractions include steel and iron structures such as underground utilities, power lines, buildings, and iron ore deposits. Sometimes these local attractions greatly alter compass readings.

LOCATION AND DIRECTION ON A QUADRANGLE MAP

Location and direction can be seen on a quadrangle map such as the one shown in Figure 3–12. This is the SW portion of the Mount Hood North Quadrangle, which shows part of the Mount Hood Wilderness and the NW portion of Mount Hood. The geographic coordinates are shown at the lower left corner of the quadrangle. That particular lower left point is 45°22′30″ north of the equator and 121°45′ west of the prime meridian. Note the number toward the right hand side on the bottom that reads 42′30″ and note the tick mark above that number. At this point, the coordinates are 45°22′30″ north of the equator and 121°42′30″ west of the prime meridian.

Notice the north arrows below the quadrangle. The UTM (Universal Transverse Mercator) grid is 0°56′ to the east of true north, and the magnetic declination is 20°E. Also note that this is based on the 1980 magnetic north.

The UTM grid is shown in groups of 1000 meters, or 1 kilometer. At the bottom of the quadrangle in Figure 3–12, the UTM grid is shown as 599, 600, 601, and 602. Along the left side, the UTM grid is shown as 5031, 5030, 5029, 5028, 5027, and 5026. The

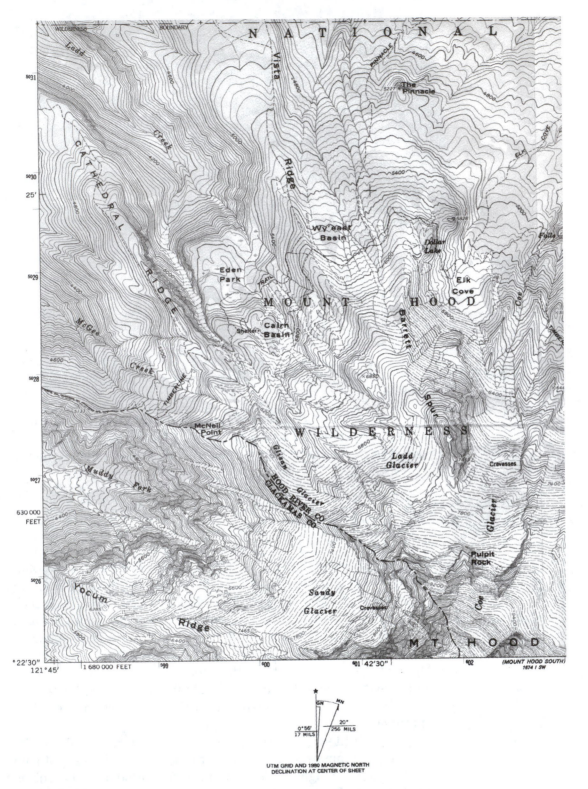

FIGURE 3–12. A portion of a quadrangle map providing location and direction.

north/south UTM (on the left side of the quadrangle) indicates the number of kilometers to the equator. Since one kilometer is equal to approximately 0.62 miles, you can conclude that the equator is about 3116.12 miles south of Yocum Ridge. (Yokum Ridge is located at the north UTM of 5026; and 5026 multiplied by 0.62 = 3116.12)

The State Plane Coordinates (SPC) are given in feet. Along the bottom near the lower left corner in Figure 3–12 is the notation "1 680 000 FEET". The SPC is shown in groups of 10,000 feet. The tick mark to the right of the notation 42′30″ is the SPC mark, and is therefore 10,000 feet from the mark of 1 680 000 FEET, or approximately 1.89 miles.

MAP GEOMETRY

The information that you have learned concerning longitude, latitude, azimuth, and bearing can be put to use in the construction of plats or plots of land. Plots of property are drawn with border lines showing the starting point for location, bearing for direction, and dimensions for size. The following discussion shows how these three components are put together to form the boundaries for a plot.

Polygon

In plotting a traverse you remember that a closed traverse is a polygon. A four-sided polygon closes if all included angles equal 360°, as in Figure 3–13. The total number of included angle degrees in polygons with more than four sides can be calculated using the formula:

$$n - 2 \times 180° = \text{Total Degrees}$$

where n = number of sides. For example, the total included angle degrees of an eight-sided polygon is calculated like this:

$$n - 2 \times 180° = \text{Total Degrees}$$
$$8 - 2 \times 180° =$$
$$6 \times 180° = 1080°$$

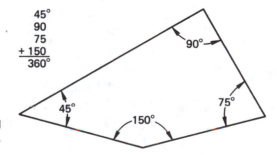

$$
\begin{array}{r}
45° \\
90 \\
75 \\
+\ 150 \\
\hline
360°
\end{array}
$$

FIGURE 3–13. All angles of a four-sided polygon will equal 360° when added.

Intersecting Lines

When two lines intersect, the opposite angles are equal. In Figure 3–14, angle $X = X$ and $Y = Y$.

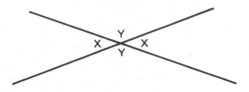

FIGURE 3–14. Opposite angles of intersecting lines are equal.

Plotting a Traverse

The foregoing concepts become important when you attempt to plot a traverse without all bearings given.

You may find a situation similar to this example. Given angles A, B, C, and D, plot the traverse and determine the bearings; be sure you understand what a bearing is. Make your calculations in a clockwise direction beginning at the point of beginning (POB). Look at Figure 3–15.

Line	Dist.	Angle	Included degrees
AB	170.1'	DAB	69°10'
BC	131.2'	ABC	110°38'
CD	173.2'	BCD	111°49'
DA	255.0'	CDA	68°23'

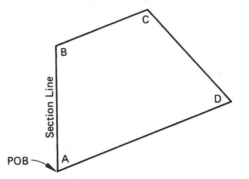

FIGURE 3–15. Example of a typical traverse. Included angles and distances given.

1. Let us begin by determining the bearing of one line at a time, first line AB. The bearing of line AB is due north. You can reason this because line AB is on the section line that is parallel to the north–south line.

2. Next, determine the bearing of line BC as shown in Figure 3–16. The bearing is in the northeast quadrant. The included angle ABC is subtracted from 180°, and the bearing is N69°22'E.

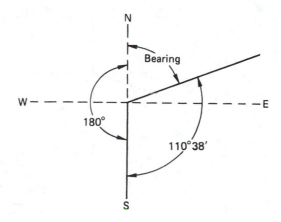

FIGURE **3–16.** Calculating bearings: line
BC.

3. Now, calculate the bearing of *CD* as shown in Figure 3–17. We first must consider everything that we know.

   a. Angle *BCD* = 111°49′.

   b. Angle *X* = 69°22′, the bearing of line *BC* that you just calculated.

   c. Angle *X* = angle *X*, because if two parallel lines are cut by another line, the exterior–interior angles on the same side of the line are equal.

   d. Angle *X* = angle *Y*, because when one straight line intersects another, the opposite angles are equal.

   e. Now, you conclude that *Y* = 69°22′, so the bearing is:

$$
\begin{array}{r}
111°49′ \\
- \quad 69°22′ \\
\hline
42°27′
\end{array}
$$

The bearing is in the southeast quadrant, so it reads S42°27′E.

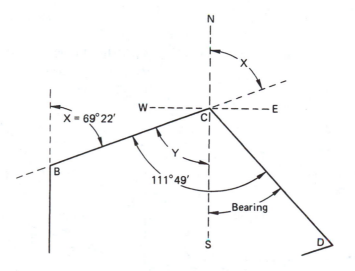

FIGURE **3–17.** Calculating bearings: line
CD.

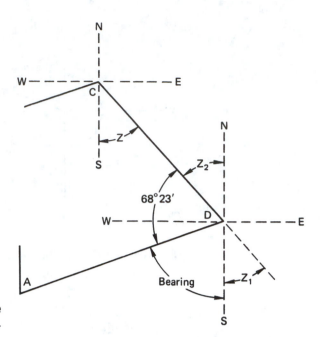

FIGURE 3–18. Calculating bearings: line *DA*.

4. Finally, determine the bearing of *DA*, and you are back to the POB as shown in Figure 3–18.

You know from the previous part that angle $Z = 42°27'$. You also know from elementary geometry that $Z = Z_1 = Z_2$. Calculate the angle between north and line *DA* like this:

$$\begin{array}{r} 68°23' \\ +\ \underline{42°27'\ (Z_2)} \\ 110°50' \end{array}$$

Now, calculate the bearing of *DA* like this:

$$\begin{array}{r} 179°60' \\ +\underline{110°50'} \\ \text{Bearing} \quad 69°10' \end{array}$$

With this bearing in the southwest quadrant, the reading is S69°10'W. Your complete plat layout should appear as shown in Figure 3–19.

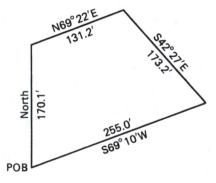

N69°22'E
131.2'

S42°27'E
173.2'

North
170.1'

255.0'
S69°10'W

POB

FIGURE 3–19. Bearings and distances shown on a plat.

Plotting a Traverse with CAD

Many of the manual calculations are done for you if you are plotting a traverse with a CAD system. However, it is advisable that you first completely understand the concepts previously discussed before you rely on a computer system to do work for you. By being initially knowledgeable of the likely outcome of a survey or a traverse, you will be more likely to detect errors even if the computer system is able to process erroneous information.

Coordinate geometry (COGO) is a means used in CAD systems for laying out survey data. The methods for laying out this data include setting individual point coordinates and setting lines and arcs by distance and bearing. Labeling survey points, lines, and curves with lengths and bearings is often an option with COGO packages.

The typical traverse with included angles and distances shown in Figure 3–15 can be easily accomplished with COGO. The information can be inserted into the computer system as is, from existing survey information, or it can be entered by keyboard.

Plotting a Closed Traverse

Plotting a traverse is a method of checking the accuracy of a survey. This consists of a series of lines and angles from the survey. Starting at a given point, these angles and lines are plotted to form a closed polygon. When you, in fact, do close the polygon, you know that the traverse is accurate, and you have what is called a *closed traverse*. Figure 3–20 is a surveyor's rough sketch of a surveyed plat. The sketch includes all line lengths and bearings. You may be given azimuths or interior angles; can you convert to bearings? An open traverse may occur when the ends do not intend to close.

Another way to present the information is in a plotting table such as the one shown in Table 3–2. If you are given a plotting table, the information may not be on the sketch.

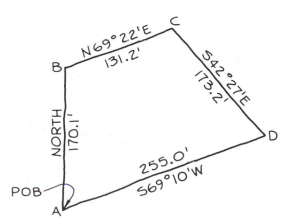

FIGURE 3–20. Rough sketch of a plat.

TABLE **3–2.** Plotting table

Line	Bearing	Distance	Angle
AB	North	170.1′	69°10′
BC	N69°22′E	131.2′	110°38′
CD	S42°27′E	173.2′	111°49′
DA	S69°10′W	225.0′	68°23′

Error of Closure

When a surveyor expects to get a closed traverse and does not quite make it, he or she can cure the problem using the *error of closure* method. As a drafter, you may also have an error due to the small inaccuracies involved in layout. These errors may result in a plot that does not quite close; one such plot is shown in Figure 3–21.

Here is what you do to make the correction:

1. Lay out the perimeter of the plot along a straight line and label the corners, as in Figure 3–22.
2. Extend the error of closure above point A'.
3. Connect a line from A'' to A.
4. Now connect perpendicular lines from points B, C, and D to line A–A''.
5. You have the distance that you can allow at each point to compensate for the problem. Notice how this is done in Figure 3–23.

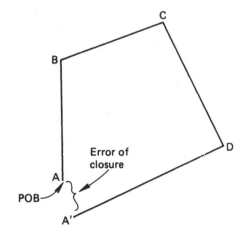

FIGURE 3–21. Error of closure.

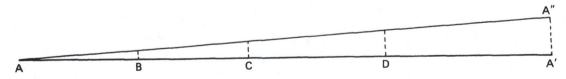

FIGURE 3–22. Correcting the error of closure.

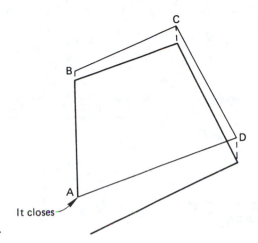

FIGURE **3-23.** Closed traverse.

PLOTTING PROPERTY USING LATITUDES AND DEPARTURES

Plotting a closed traverse is often done by calculating latitude and departure. The *latitude* of a property line relates to the distance the line extends in a north or south direction. A *positive latitude* extends in a northerly direction, while a *negative latitude* runs southerly. See Figure 3–24.

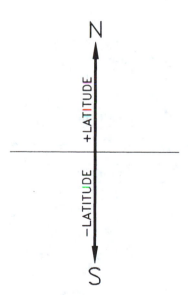

FIGURE **3-24.** Positive and negative latitude.

The *departure* of a property line is the distance the property line extends in an east-west direction. Lines running easterly have a *positive departure*, while *negative departures* run westerly as shown in Figure 3–25.

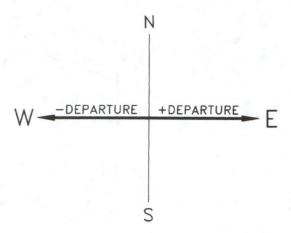

FIGURE **3–25.** Positive and negative departures.

A right triangle is created by a property line and its latitude and departure. The latitude of a property line is the vertical side of the right triangle, and the departure is the horizontal side as illustrated in Figure 3–26.

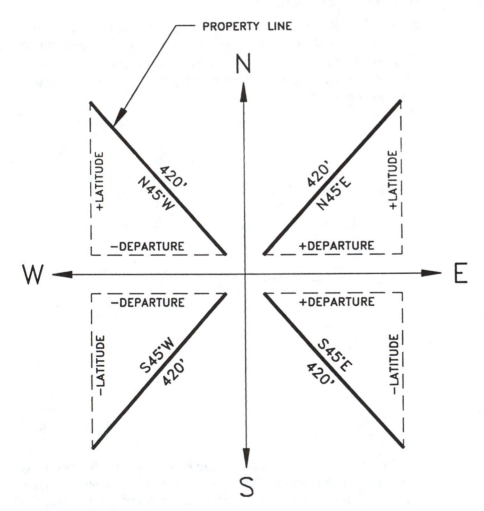

FIGURE **3–26.** A right triangle created by property lines with latitude and departure.

The latitude and departure values are calculated as follows:

$$\text{Latitude} = \text{Distance} \times \cos \text{Bearing}$$
$$\text{Departure} = \text{Distance} \times \sin \text{Bearing}$$

The latitude and departure of the N45°E property line shown in Figure 3–26 is calculated like this:

$$\text{Latitude} = 420' \times \cos 45° = 420' \times 0.70711 = 296.9862$$
$$\text{Departure} = 420' \times \sin 45° = 420' \times 0.70711 = 296.9862$$

The latitude and departure of the N45°E property line are the same because the bearing is a 45° angle. These values would be different for property lines with other bearings. With regard to the S45°W bearing, the latitude is −296.9862 and the departure is −296.9862.

It is a good idea to set up a table when working with latitude and departure calculations. This helps you keep the information accurately organized. Look at Figure 3–27 and notice the property lines and resulting latitude and departure calculations placed on the drawing and in a table.

The importance of latitudes and departures lies in the plotting of a closed traverse. When a traverse is closed, the sum of the latitudes and departures is zero. Latitudes and departures are referred to as *balanced* when their sums equal zero. When the sum of the latitudes and departures does not equal zero, the resulting amount is the error of closure and the data is called *unbalanced*. Unbalanced latitudes and departures require that slight changes be made to the latitude and departure of each property line in order to close the traverse and balance the calculations at zero.

Computer programs are available that process the information about a plot based on the bearings and distances of each property line and automatically calculate the latitudes, departures, and azimuths. If the resulting data is unbalanced, automatic adjustments are made in property lines to balance the latitudes and departures and to close the traverse. A set of calculations of this type is shown in Figure 3–28.

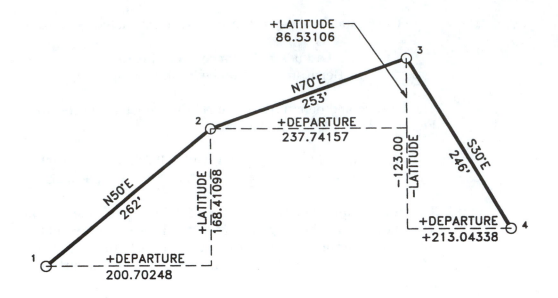

COURSE	BEARING	DISTANCE	COS	SIN	LATITUDE	DEPARTURE
1-2	N50°E	262'	.64279	.76604	168.41098	200.70248
2-3	N70°E	253'	.34202	.93969	86.53106	237.74157
3-4	S30°E	246'	.86603	.5000	-123.0000	213.04338

FIGURE 3-27. Setting up a table for latitude and departure calculations.

Traverse Computation
Problem Number 1: Polygon traverse.

Course	Length	Azimuth	Cos(Az)	Sin(Az)
1-2	284.800	21 3 0.0	+0.93326741	+0.35918251
2-3	210.000	93 35 0.0	−0.06250017	+0.99804503
3-4	240.150	93 3 0.0	−0.05320745	+0.99858356
4-5	278.750	201 3 0.0	−0.93326730	−0.35918263
5-1	452.000	272 34 0.0	+0.04478198	−0.99899679

1,465.700 = total distance

Unbalanced		Balanced			Coordinates	
Lat	Dep	Lat	Dep	Point	X	Y
+265.795	+102.295	+265.797	+102.290	1	0.000	0.000
−13.125	+209.589	−13.123	+209.586	2	102.290	265.797
−12.778	+239.810	−12.775	+239.806	3	311.876	252.675
−260.148	−100.122	−260.145	−100.127	4	551.682	239.899
+20.241	−451.547	+20.246	−451.554	5	451.554	−20.246
−0.015	+0.026					

Linear error of closure = .029 ft.
Precision = 1 in 49117
Area = 2.7677 acres

FIGURE 3-28. Plot information processed by a computer automatically calculates and balances latitudes, departures, and azimuths.

TEST

Part I

Multiple choice: Circle the response that best finishes each statement.

3-1   The line that represents zero degrees longitude is also called:

a. The International Date Line
b. The prime meridian
c. The equator
d. Greenwich, England

3-2   Lines of latitude:

a. Are measured as an angular distance from the point at the center of the earth
b. Connect the north and south poles
c. Are called parallels
d. Are approximately the same distance apart

3-3   45°26′14″ + 15°10′52″ =

a. 60°36′6″
b. 60°37′66″
c. 60°37′6″
d. 61°13′6″

3-4   A direction measured clockwise from a given zero is called:

a. A bearing
b. An azimuth
c. A magnetic declination
d. A local attraction

3-5   A direction measured clockwise or counterclockwise with respect to quadrants of a compass is called:

a. A grid azimuth
b. An azimuth
c. A bearing
d. A magnetic declination

3-6   Refer to Figure 3–12. What is the distance from Elk Cove to the equator?

a. 5,020,000 meters
b. 5,029,000 miles
c. 5029 kilometers
d. 8,111.3 miles

Part II

3-1 Give the azimuths of the following lines. North is zero.

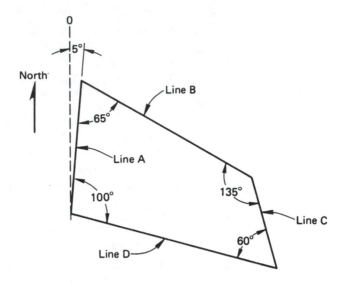

Line *A* _____

Line *B* _____

Line *C* _____

Line *D* _____

3-2 Give the bearings of the lines featured in the question above.

Line *A* _____

Line *B* _____

Line *C* _____

Line *D* _____

Part III

True or false: Circle the "T" if the statement is true or the "F" if the statement is false. Reword all false statements so that the meaning is true.

3-1 T F Magnetic azimuths may differ from true azimuth by several degrees.

3-2 T F The horizontal angle between the magnetic meridian and the true meridian is called the magnetic declination.

3-3 T F Magnetic azimuth + magnetic declination = true azimuth.

3-4 T F The magnetic declination does not change.

3-5 T F The included angles of a four-sided polygon equal 360°.

3-6 T F When drawing a plat based on angular information, you must convert bearings to azimuths or interior angles.

Part IV

Given the following groups of information, determine the distances on the earth's surface. Use your knowledge of lengths of a degree of longitude and latitude. Make sketches of the earth showing the points given to help you establish location, and show calculations where appropriate.

3-1 The distance between a point at 45° north latitude, 20° west longitude and a point 30° north latitude, 20° west longitude.

3-2 The distance between a point at 16° north latitude, 60° east longitude and a point at 28° south latitude, 60° east longitude.

3-3 The distance between a point at 50° north latitude, 80° west longitude and a point at 50° north latitude, 20° west longitude.

3-4   Calculate the length of a degree of longitude at 45° north latitude.

3-5   Calculate the length of a degree of longitude at 75° south latitude.

3-6   Calculate the true azimuth given the following information:
East magnetic azimuth = 32°
Magnetic declination = 15° E

3-7   Calculate the true azimuth given the following information:
East magnetic azimuth = 46°
Magnetic declination = 5° W

3-8 Calculate the true azimuth given the following information:
West magnetic azimuth = 18°
Magnetic declination = 7° E

3-9 Calculate the true azimuth given the following information:
East magnetic azimuth = 27°
Magnetic declination = 12° W

3-10 Calculate the true azimuth given the following information:
East magnetic azimuth = 19°45'12"
Magnetic declination = 4° 50'39" E

3-11 Find the distance between a point at 10° north latitude, 80° west longitude and a point at 10° north latitude, 20° west longitude.

3-12 Determine the distance between a point at 72° north latitude, 28° west longitude to a point at 18° north latitude, 28° west longitude to a point at 18° north latitude, 42° west longitude.

PROBLEMS

P3-1   Tape Figure P3–1 to your drawing board. With your drafting machine protractor or hand-held protractor, determine the azimuth and bearing for each line. Place your answer in the table provided. You will be evaluated on accuracy to ± 15′ and on the neatness of your lettering in the table.

Line	Azimuth	Bearings
A		
B		
C		
D		
E		
F		
G		
H		
I		

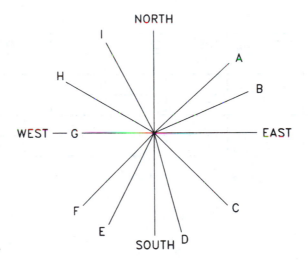

FIGURE **P3–1.**

P3-2 Given the partial traverse in Figure P3–2, place the following items in the table:

Bearing of each property line (course)
Distance of each course
Cosine of each bearing angle
Sine of each bearing angle
Latitude of each course
Departure of each course

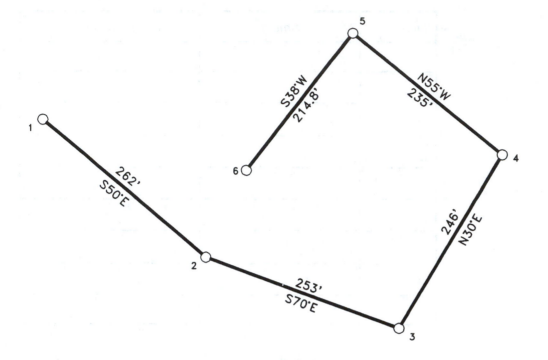

COURSE	BEARING	DISTANCE	COS	SIN	LATITUDE	DEPARTURE
1–2						
2–3						
3–4						
4–5						
5–6						

FIGURE **P3–2.**

P3-3 Given the traverse in Figure P3–3, place the following items in the table:
Bearing of each property line (course)
Distance of each course
Cosine of each bearing angle
Sine of each bearing angle
Latitude of each course
Departure of each course

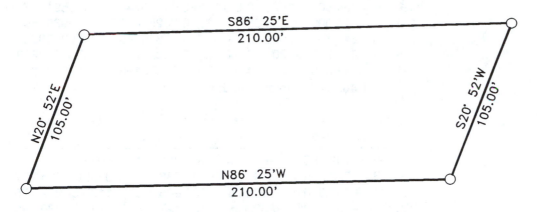

COURSE	BEARING	DISTANCE	COS	SIN	LATITUDE	DEPARTURE
1–2						
2–3						
3–4						
4–1						

FIGURE **P3–3.**

P3-4 Draw a traverse from the calculations and information given in Figure P3–4.

Traverse Computation
Problem Number 1: Polygon traverse.

Course	Length	Azimuth	Cos(Az)	Sin(Az)
1-2	284.800	21 3 0.0	+0.93326741	+0.35918251
2-3	210.000	93 35 0.0	−0.06250017	+0.99804503
3-4	240.150	93 3 0.0	−0.05320745	+0.99858356
4-5	278.750	201 3 0.0	−0.93326730	−0.35918263
5-1	452.000	272 34 0.0	+0.04478198	−0.99899679

1,465.700 = total distance

Unbalanced		Balanced			Coordinates	
Lat	Dep	Lat	Dep	Point	X	Y
+265.795	+102.295	+265.797	+102.290	1	0.000	0.000
−13.125	+209.589	−13.123	+209.586	2	102.290	265.797
−12.778	+239.810	−12.775	+239.806	3	311.876	252.675
−260.148	−100.122	−260.145	−100.127	4	551.682	239.899
+20.241	−451.547	+20.246	−451.554	5	451.554	−20.246
−0.015	+0.026					

Linear error of closure = .029 ft.
Precision = 1 in 49117
Area = 2.7677 acres

FIGURE **P3–4.**

P3-5 Figure P3–5a shows a small subdivision drawn with a CAD system. Given the northing and easting information in the table in Figure P3–5b, use your CAD system to plot the subdivision boundary. Identify the length (to one-hundredth of a foot) and bearing of each property line and place the information in the table. Plot the traverse at a scale of 1 in. = 50 ft on a B-size sheet of paper and place the following items in your plot:

Bearing of each line (course)
Distance of each course
Small circles at intersections of courses. (See Figure P3–2.)
Point number at end of each course. (See how the course is numbered in Figure P3–2.)
POB
Scale
North arrow
Title block: Erika Subdivision, Quincey Lane at Christopher Blvd., Jean, Wisconsin

FIGURE P3–5a. Graphic supplied courtesy of Environmental Systems Research Institute, Inc.

POINT NO.	NORTHING	EASTING	LENGTH	BEARING
1 (POB)	71394.4238	176534.3439		
2	71368.0963	176968.3491		
3	71127.4345	176950.6149		
4	71028.9085	176905.4732		
5	70969.1469	176571.7470		
6	71143.5863	176510.4833		

FIGURE P3–5b.

P3-6 Refer to Figure 3–12. Using the coordinates found on that quadrangle, mark the approximate location of Mount Hood on the map shown in Figure P3–6. Draw the approximate magnetic declination line of the quadrangle on Figure P3–6.

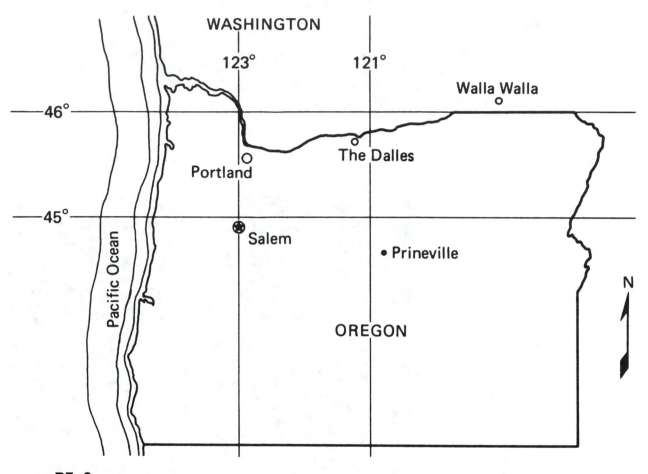

FIGURE **P3–6.**

CHAPTER 4

Mapping Scales

This chapter explains the purpose and types of map scales. Illustrations are provided to show how scales affect the map that is being represented.

The topics covered include:

- Numerical scale
- Graphic scale
- Verbal scale
- Scale conversion
- Use of the engineer's scale
- Scales for CAD

Map scales vary as to purpose, size, and desired detail. For example, if a map of the State of New York were to be painted to scale on a large, blank wall of a building, the scale would be different than if the New York map were drawn to scale on an 11-by 17-in. sheet of paper. Also, a map scale varies according to the area of the earth's surface covered. A map of your school drawn on this page will have a different scale than if it were a map of the United States.

A simple definition of a map is a representation of a portion of the earth's surface reduced to a small size. The map's scale aids you in estimating distances, and gives you an idea of what to expect in the way of detail. The map scale shows the relationship between the measurements of the features shown on the map compared to the same features on the earth's surface.

There are three methods of expressing scale:

1. Numerical scale or representative fraction
2. Graphic scale
3. Verbal scale

NUMERICAL SCALE

The numerical scale gives the proportion between the length of a line on a map and the corresponding length on the earth's surface. This proportion is known as the *representative fraction* (RF). The first number is a single unit of measure equal to the distance on the map = 1. The second number is the same distance on the earth using the same units of measure. It can be written in two ways: 1/150,000 or 1:150,000.

RF (representative fraction) = distance on map/distance on earth

It is very important to note that this map scale must always refer to map and ground distances in the same unit of measure. The unit of measure most commonly used in the representative fraction is the inch. For instance, a scale of 1 in. on your map representing 1 mile on the earth's surface would be expressed as follows: RF = 1/63,360 or 1:63,360. (One mile has 63,360 in.)

Remember, the numerator is always 1, and represents the map. The denominator is always greater, because it represents the ground.

To calculate the distance between two points on a map, you must multiply the measured distance in inches by the denominator of the RF. Let us presume that you have measured $2\frac{1}{4}$ in. between points A and B on a map drawn to a scale of 1/100,000. The real distance between points A and B would be: distance = $2\frac{1}{4}$ × 100,000 = 225,000 in. You can convert this to feet or miles if you wish. If the map were calibrated in metrics, 1:100,000 would mean 1 cm = 100,000 cm = 1 km.

Small-Scale and Special Maps

Information compiled in a pamphlet titled *Tools for Planning* by the U.S. Department of Interior, Geological Survey, indicates that several series of small-scale and special maps are published to meet modern requirements. These include:

1. *1:250,000 scale* (1 in. on the map represents about 4 miles on the ground)—for regional planning and as topographic bases for other types of maps.

2. *State maps*, at 1 : 500,000 scale (1 in. represents 8 miles)—for use as wall maps and for statewide planning.

3. *Shaded-relief maps*, at various scales—for park management and development and as tourist guides.

4. *Antarctic topographic maps*, at various scales—for scientific research.

5. *International map of the world* (IMW), at 1 : 1,000,000 scale—for broad geographic studies.

Figure 4–1 provides a graphic example of maps at different scales.

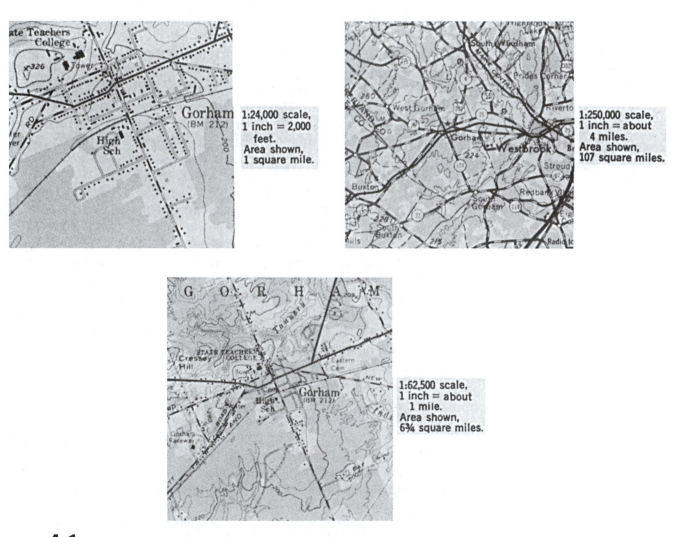

FIGURE 4–1. Reproduced from "Topographic Maps" by permission of the U.S. Geological Survey.

This kind of map scale is like a small ruler in the legend or margin of the map. The divisions on the graphic scale represent increments of measure easily applied to the map.

Ordinarily, graphic scales begin at zero, but many have an extension to the left of the zero. This enables you to determine distances less than the major unit of the scale. A graphic scale uses relatively even units of measure, such as 500 or 1000. Figure 4–2 illustrates three graphic scales, each calibrated differently.

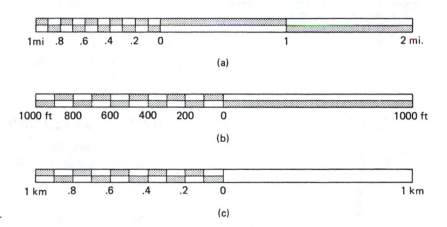

FIGURE **4–2.** Sample graphic scales.

Suppose you wish to measure the distance down the road between points *A* and *B* in the simple map in Figure 4–3. Just apply a straight-edged strip of plain paper on the map along the distance between the two points and make marks on the paper strip where the edge touches *A* and *B*. Then move the strip of paper to the graphic scale and measure. Point *A* is found to be 1.5 miles down the road from point *B*. Figure 4–3 also illustrates how the piece of paper is used to mark the distance on the map and then determine the distance measured.

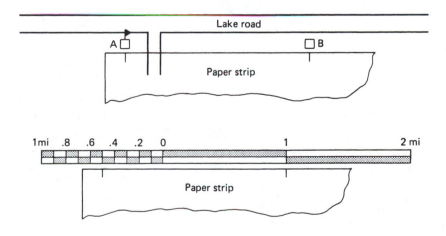

FIGURE **4–3.** Measuring with a graphic scale.

VERBAL SCALE

Verbal scale is usually expressed in the number of inches to the mile. Usually, the mileage is rounded. For example, 1 in. on an RF scale of 1 : 1,000,000 equals 15.78 mi. This distance expressed in verbal scale would round off the 15.78 to 16. So the verbal scale would be 1 in. = 16 mi. This is a close approximation for discussion purposes. The verbal scale is not meant to be accurate, only a rough estimate. Use the RF scale for accurate measurements.

SCALE CONVERSION

In map drafting or map reading it is often necessary to convert from one scale to another. You may have to convert an RF scale to a graphic scale for display on the map.

Convert a Representative Fraction to a Graphic Scale

Example: Given a representative scale of 1 : 300,000, find the graphic equivalent. Remember that the 1 equals 1 in. on the map and the 300,000 equals 300,000 in. on the earth's surface. We must first determine the number of miles that are represented by the 300,000 in. There are 63,360 in. in a mile, so

$$\frac{300,000 \text{ in}}{63,360 \text{ in.}} = 4.73 \text{ miles}$$

This scale is not practical because it is difficult to imagine odd measurements such as 4.73 miles as shown in Figure 4–4.

FIGURE **4–4.** This graphic scale is not practical.

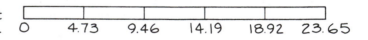

Instead, graphic scales are drawn using units like 0-1-2-3-4 or 0-2-4-6 or 0-5-10-15-20. To do this, it is necessary to find out how many inches are used to show any even number larger than the 4.73 miles previously determined. Let us choose 10 mi. This would be set up in a ratio:

$$\frac{4.73 \text{ miles}}{1 \text{ in.}} = \frac{10 \text{ miles}}{\text{(unknown no. of inches)}}$$

Use cross-multiplication to establish this algebraic formula:

$$4.73x = 10$$
$$x = 2.11 \text{ in.}$$

In 2.11 in. you have 10 miles. Then it is easy to lay off a line 2.11 in. long and divide it into 10 equal parts using the method of parallel lines, shown in Figure 4–5.

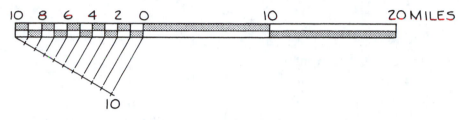

FIGURE 4–5. Correct graphic scale with one segment divided into 10 equal spaces.

Convert a Graphic Scale to a Representative Fraction

Example: 1 in. on the graphic scale equals 5 mi (1 mi = 63,360 in.).

$$5 \times 63,360 = 316,800$$
$$RF = 1:316,800$$

Determine the representative fraction for a map with no scale.

The representative fraction can be determined by measuring on a straight meridian line the distance for the map's 1° of latitude. The actual average distance in 1° of latitude is 69 miles. So if the number of inches between 1° of latitude equals 2.3, the following formula will work to solve the RF:

$$RF = \frac{\text{miles in 1° of latitude} \times 63,360 \text{ in.}}{\text{inches in 1° of latitude on the map}}$$
$$= \frac{69 \times 63,360}{2.3}$$
$$= 1:1,900,800$$

ENGINEER'S SCALE

The maps that you draw will require a scale so that the reader will be able to interpret distances accurately. Different methods of expressing map scales have been discussed. Now, you need a tool to use that will allow you to draw a map at a scale that you select. Commonly, maps are drawn using a scale that is made up in mul-

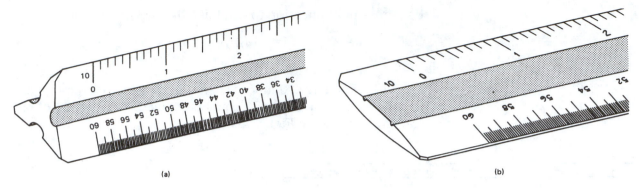

(a)   (b)

FIGURE 4–6. (a) A typical triangular engineer's scale. (b) Flat engineer's scale.

tiples of 10. For example, 1 in. = 10 ft, 100 ft, 1000 ft, 10,000 ft, and so on. The tool that is often used by the drafter is a civil engineer's scale. Figure 4–6 shows an example of a typical triangular and flat engineer's scale.

There are six engineer's scales found on the triangular engineer's scale. Each scale is a multiple of 10, and may be used to calibrate a drawing in any units, such as feet, meters, miles, or tenths of any typical unit.

This list shows the six possible scales:

10 scale: 1 in. = 1.0 ft; 1 in. = 10.0 ft; 1 in. = 100.0 ft; etc.
20 scale: 1 in. = 2.0 ft; 1 in. = 20.0 ft; 1 in. = 200.0 ft; etc.
30 scale: 1 in. = 3.0 ft; 1 in. = 30.0 ft; 1 in. = 300.0 ft; etc.
40 scale: 1 in. = 4.0 ft; 1 in. = 40.0 ft; 1 in. = 400.0 ft; etc.
50 scale: 1 in. = 5.0 ft; 1 in. = 50.0 ft; 1 in. = 500.0 ft; etc.
60 scale: 1 in. = 6.0 ft; 1 in. = 60.0 ft; 1 in. = 600.0 ft; etc.

Along the margin of the engineer's scale, you will see a 10 on one edge. This 10 represents the 10 scale. Another edge will have a 20 in the margin representing the 20 scale. This same situation holds true for the 30, 40, 50, and 60 scales. Figure 4–7 is a close-up of the margin on an engineer's scale.

Measurements are easy to multiply and divide since they are given in decimals rather than in feet and inches. Figure 4–8 pro-

FIGURE 4–7. Margin of triangular engineer's scale.

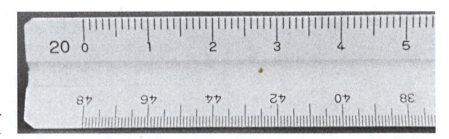

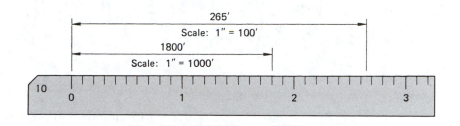

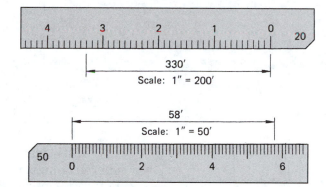

FIGURE **4–8.** Three examples of margins and scales on an engineer's scale.

vides some excellent illustrations of how the engineer's scale may be used to measure distances.

METRIC SCALES

The unit of measure commonly used in metric scales is the millimeter, which is based on the International System of Units (SI). Metric scales are abbreviated as follows:

Unit	Abbreviation
millimeter	mm
centimeter	cm
decimeter	dm
meter	m
dekameter	dam
hectometer	hm
kilometer	km

Metric scales are convenient to work with because they are calibrated in units of 10, which makes metric-to-metric conversions easy as follows:

$$10 \text{ mm} = 1 \text{ cm}$$
$$10 \text{ cm} = 1 \text{ dm}$$
$$10 \text{ dm} = 1 \text{ m}$$
$$10 \text{ m} = 1 \text{ dam}$$
$$10 \text{ dam} = 1 \text{ hm}$$
$$10 \text{ hm} = 1 \text{ km}$$

Common metric-to-U.S. equivalents are as follows:

$$1 \text{ mm} = 0.03937 \text{ in.}$$
$$1 \text{ cm} = 0.3937 \text{ in.}$$
$$1 \text{ m} = 39.37 \text{ in.}$$
$$1 \text{ km} = 0.624 \text{ mi}$$

Common U.S.-metric equivalents are as follows:

$$1 \text{ mi} = 1.6093 \text{ kilometers} = 1609.3 \text{ meters}$$
$$1 \text{ yd} = 0.9144 \text{ m} = 914.4 \text{ mm}$$
$$1 \text{ ft} = 0.3048 \text{ m} = 304.8 \text{ mm}$$
$$1 \text{ in.} = 0.0254 \text{ m} = 25.4 \text{ mm}$$

The most commonly used inch-to-metric conversion formula is:

$$\text{In.} \times 25.4 = \text{mm}$$

When placing metric dimensions on a drawing, the metric abbreviation is usually omitted. A general note specifying METRIC, or UNLESS OTHERWISE SPECIFIED, ALL DIMENSIONS ARE IN MILLIMETERS is placed on the drawing.

Metric scales may be enlarged or reduced as needed for the desired drawing scale, because any scale is a multiple of 10. For example, enlarging a scale such as the 1 : 5 scale in multiples of 10 makes the 100 division equal to 1000, thus creating a 1 : 50 scale.

The following displays common map scales and their U.S. custom and metric equivalents:

Map Scale	U.S. Custom	Metric
1 : 10,000	0.158 mile (mi)	0.1 km
1 : 25,000	0.395 mi	0.25 km
1 : 50,000	0.789 mi	0.5 km
1 : 75,000	1.18 mi	0.75 km
1 : 100,000	1.58 mi	1 km
1 : 250,000	3.95 mi	2.5 km
1 : 500,000	7.89 mi	5 km
1 : 1,000,000	15.78 mi	10 km

Common metric scales are shown in Figure 4–9.

CAD SCALES

Scaling CAD drawings is somewhat different from scaling manual drawings. CAD drawings are most often drawn using *real world coordinates*. This means that, if a lot line is 100.00 ft long the line

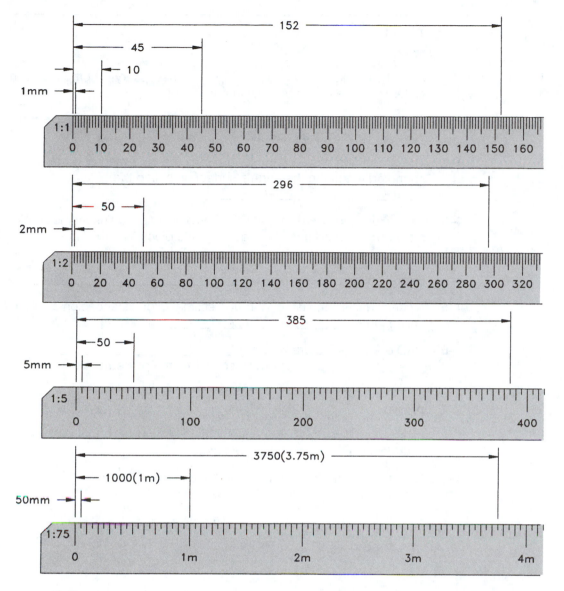

FIGURE 4-9.

is actually drawn in CAD at exactly 100.00 ft instead of taking into account a scale factor. The drawing scale is therefore 1 : 1. The scale factor is then brought into effect when the drawing is ready to be plotted or printed. For example, if the final plotted drawing is to be 1 in. = 50 ft, that information is given when the CAD drawing is sent to the plotter, and the CAD system automatically scales the entire drawing down to that size.

This is necessary especially when working with digital topography maps as overlays. The topography maps are also drawn using real world coordinates, and, to overlay them successfully with the CAD drawing, the scales need to be exactly the same.

TEST

Part I

Fill in the blanks below with the responses necessary to complete each statement. Use your best freehand lettering.

4–1 A _____ is a representation of a portion of the earth's surface reduced to a small size.

4–2 The proportion between the length of a line on a map and the corresponding length on the earth's surface is called the _____ _____.

4–3 A scale used to approximate distances on a map, that usually looks like a small ruler found near the legend or title block, is called the _____ _____.

4–4 Representative fraction = RF = _____ / _____.

4–5 A scale that is often used for rough estimates of distances on a map is called the _____ _____.

4–6 CAD drawings are drawn using, _____ _____ _____, and the drawn scale (not the plotted scale) is. _____ : _____.

Part II

Make the following scale conversions. Be neat and accurate. Show your calculations in the space provided.

4–1 Convert RF = 1 : 100,000 to a graphic scale.

4–2 Convert RF = 1 : 2000 to a graphic scale.

4–3   Convert a verbal scale of 1 in. = 1 mi to a representative fraction.

4–4   Convert a verbal scale of 1 in. = 100 mi to RF.

4–5   If the number of inches between 1° of latitude on a map is measured to be 4.75 in., what is the RF of the map?

Part III

Use your civil engineer's scale to measure each line shown below. (The scale calibration is identified to the left of each line.) Place your answer to the right.

		Scale	Answer
4–1	1 in. = 10 ft	_____	_____
4–2	1 in. = 20 ft	_____	_____
4–3	1 in. = 30 ft	_____	_____
4–4	1 in. = 40 ft	_____	_____
4–5	1 in. = 50 ft	_____	_____
4–6	1 in. = 60 ft	_____	_____
4–7	1 in. = 2 ft	_____	_____
4–8	1 in. = 100 ft	_____	_____

PROBLEM

P4–1 Convert the engineer's sketch in Figure P4–1 to a formal drawing. Remember, the sketch is not accurate, so you will have to lay out all bearings and distances. Do not trace the sketch. Consider the following.

1. Use C-size vellum 17 × 22 in.
2. Use pencil.
3. Include:
   a. Title
   b. Scale 1 in. = 10 ft or 1 in. = 20 ft
   c. Graphic scale
   d. North arrow
   e. Your name and date
4. Make a print for instructor evaluation unless otherwise specified.

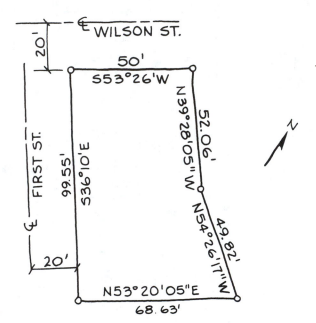

FIGURE P4–1.

CHAPTER 5

Mapping Symbols

This chapter discusses and shows examples of some of the symbols that are commonly used in civil drafting. These are only a few of the actual number of symbols that are available for your use. The purpose of the map dictates the symbols to be used. Special symbols sometimes have to be designed for a specific purpose. The symbols shown may change slightly. Symbols may include colors or may be in black and white.

The topics covered include:

1. Types of symbols
2. Symbol colors, when used
3. Special-effect symbols

Standard symbols are used in civil drafting as they are in any of the other fields of drafting. When you look at a road map, for example, you generally take it for granted that symbols on the map will help you get where you want to go. Without these symbols, the map would be useless. Each type of map may have certain symbols that are special to the intent of the map. Specific symbols are also used to help keep information on a map to a minimum so that the map remains as uncluttered as possible. Symbols are as condensed and to the point as possible to describe the objective.

TYPES OF SYMBOLS

The various symbols used in map drawings can be grouped under four types: culture, relief, water, and vegetation.

Culture

Culture symbols represent works of people. When maps are printed in color, these symbols are usually done in black. Lettering on maps that identify cultural representation is usually done in vertical caps. Variations will be found to standards that exist. The topographic map symbols shown in Figure 5–1 show several different cultural symbols.

Relief

Relief symbols are used to show the characteristics of the land. Mountains, canyons, and other land features are shown by relief. The relief may be shown by contour lines or by more descriptive means such as technical shading methods. *Contour lines* are lines on a map that represent points of the same elevation. Contour lines are usually thin lines drawn freehand, or in CAD with every fifth line being drawn thicker and broken somewhere to insert the elevation number. This number represents the elevation of the line in feet above sea level. Lettering on a map in conjunction with relief is usually done in vertical uppercase letters. Contour lines and other relief symbols are usually brown when the map is in color. Contour lines are discussed in Chapter 7.

Water

Water features represent lakes, rivers, streams, and even intermittent waters. Water features are colored blue when the map is in color. Something unique about water features is that when labeled on a map, they are done using uppercase slanted letters. For example:

MISSISSIPPI RIVER

Look at Figure 5–1.

Vegetation

Vegetation features include forests, orchards, croplands, and other types of plant life. When a map is done in color, these features are usually green. Look at Figure 5–1. If the vegetation is planted by people, such as a field of corn, some maps classify these as culture.

Provisional edition maps
New or replacement standard edition maps
Standard edition maps

CONTROL DATA AND MONUMENTS

	Standard	New/repl.	Provisional
Aerial photograph roll and frame number	Not Shown	Not Shown	3-20

Horizontal control:

Third order or better, permanent mark	Neace △	Neace △	Neace ⊕
With third order or better elevation	BM △ 148	BM △ 45.1	Pike BM 45.1
Checked spot elevation	△ 64	△ 19.5	Not Shown
Coincident with section corner	Cactus	Cactus	Cactus
Unmonumented	Not Shown	Not Shown	+

Vertical control:

Third order or better, with tablet	BM × 53	BM × 16.3	BM × 53.4
Third order or better, recoverable mark	× 394	× 120.0	× 393.6
Bench mark at found section corner	BM + 61	BM + 18.6	BM + 60.9
Spot elevation	× 17	× 5.3	× 17

Boundary monument:

With tablet	BM □ 71	BM □ 21.6	BM ⊞ 71
Without tablet	□ 562	□ 171.3	□ 562
With number and elevation	67 □ 988	87 □ 301.1	67 □ 988 USMM
U.S. mineral or location monument	▲	▲	▲

BOUNDARIES

National
State or territorial
County or equivalent
Civil township or equivalent
Incorporated-city or equivalent
Park, reservation, or monument
Small park

LAND SURVEY SYSTEMS

U.S. Public Land Survey System:
Township or range line
  Location doubtful
Section line
  Location doubtful
Found section corner; found closing corner
Witness corner; meander corner ... WC + MC

Other land surveys:
  Township or range line
  Section line
Land grant or mining claim; monument
Fence line

ROADS AND RELATED FEATURES

Primary highway
Secondary highway
Light duty road
Unimproved road
Trail
Dual highway
Dual highway with median strip
Road under construction
Underpass; overpass
Bridge
Drawbridge
Tunnel

BUILDINGS AND RELATED FEATURES

Dwelling or place of employment: small; large
School; church
Barn, warehouse, etc.: small; large
House omission tint
Racetrack
Airport
Landing strip
Well (other than water); windmill
Water tank: small; large
Other tank: small; large
Covered reservoir
Gaging station
Landmark object
Campground; picnic area
Cemetery: small; large

FIGURE 5-1. Standard topographic map symbols. Reproduced by permission of the U.S. Geological Survey.

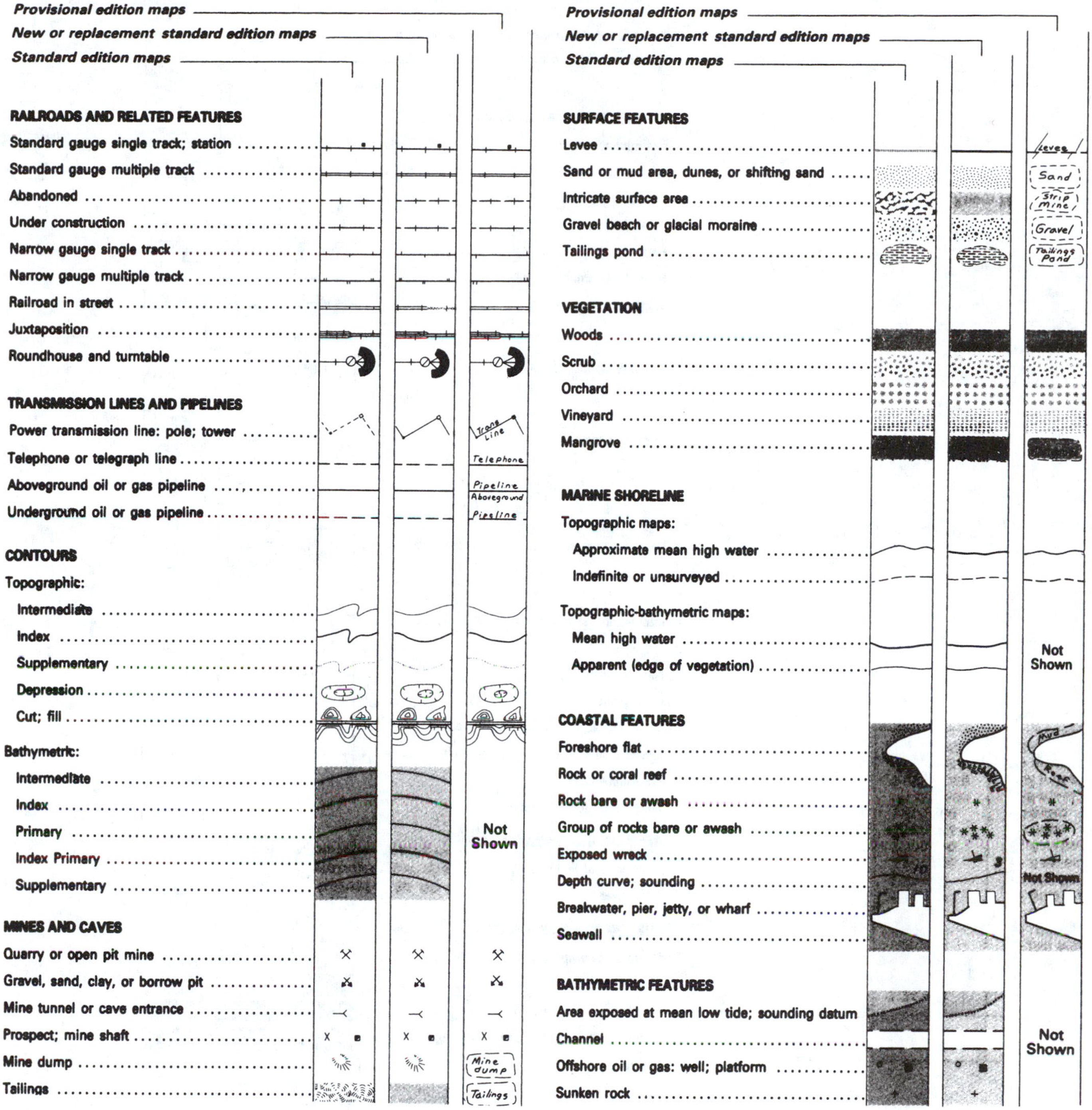
Provisional edition maps
New or replacement standard edition maps
Standard edition maps
RAILROADS AND RELATED FEATURES
Standard gauge single track; station
Standard gauge multiple track
Abandoned
Under construction
Narrow gauge single track
Narrow gauge multiple track
Railroad in street
Juxtaposition
Roundhouse and turntable
TRANSMISSION LINES AND PIPELINES
Power transmission line: pole; tower
Telephone or telegraph line
Aboveground oil or gas pipeline
Underground oil or gas pipeline
Trans Line
Telephone
Pipeline
Aboveground
Pipeline
CONTOURS
Topographic:
Intermediate
Index
Supplementary
Depression
Cut; fill
Bathymetric:
Intermediate
Index
Primary
Index Primary
Supplementary
Not Shown
MINES AND CAVES
Quarry or open pit mine
Gravel, sand, clay, or borrow pit
Mine tunnel or cave entrance
Prospect; mine shaft
Mine dump
Tailings
Mine dump
Tailings
Provisional edition maps
New or replacement standard edition maps
Standard edition maps
SURFACE FEATURES
Levee
Sand or mud area, dunes, or shifting sand
Intricate surface area
Gravel beach or glacial moraine
Tailings pond
Levee
Sand
Strip Mine
Gravel
Tailings Pond
VEGETATION
Woods
Scrub
Orchard
Vineyard
Mangrove
MARINE SHORELINE
Topographic maps:
Approximate mean high water
Indefinite or unsurveyed
Topographic-bathymetric maps:
Mean high water
Apparent (edge of vegetation)
Not Shown
COASTAL FEATURES
Foreshore flat
Rock or coral reef
Rock bare or awash
Group of rocks bare or awash
Exposed wreck
Depth curve; sounding
Breakwater, pier, jetty, or wharf
Seawall
Mud
Reef
Not Shown
BATHYMETRIC FEATURES
Area exposed at mean low tide; sounding datum
Channel
Offshore oil or gas: well; platform
Sunken rock
Not Shown

FIGURE 5–1. *(continued)*

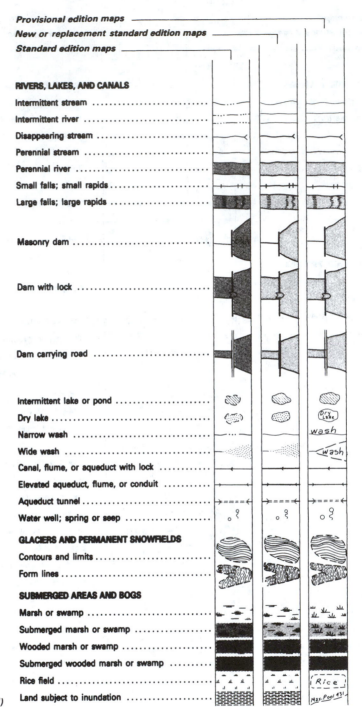

Provisional edition maps
New or replacement standard edition maps
Standard edition maps

RIVERS, LAKES, AND CANALS

Intermittent stream ....

Intermittent river ....

Disappearing stream ....

Perennial stream ....

Perennial river ....

Small falls; small rapids..................

Large falls; large rapids ....

Masonry dam ....

Dam with lock ....

Dam carrying road ....

Intermittent lake or pond ....

Dry lake ....

Narrow wash ....

Wide wash ....

Canal, flume, or aqueduct with lock ....

Elevated aqueduct, flume, or conduit ....

Aqueduct tunnel ....

Water well; spring or seep ....

GLACIERS AND PERMANENT SNOWFIELDS

Contours and limits ....

Form lines ....

SUBMERGED AREAS AND BOGS

Marsh or swamp ....

Submerged marsh or swamp ....

Wooded marsh or swamp ....

Submerged wooded marsh or swamp ....

Rice field ....

Land subject to inundation ....

FIGURE **5–1.** *(continued)*

Symbols have been made by various governmental agencies for their particular maps. Also, agencies of private industry may have their own mapping symbols. Some examples are:

Governmental

1. Federal Board of Surveys and Maps
2. National Oceanic and Atmospheric Administration (formerly U.S. Coast and Geodetic Survey)
3. U.S. Geological Survey
4. U.S. Forest Service
5. Army Map Service

Private Industry

1. American Railway Engineering Association
2. American Consulting Engineers Council

SPECIAL TECHNIQUES

Other methods are often used to show symbols or to represent specific concepts on a map. For example, it was discussed earlier that relief is sometimes shown using graphic techniques. This is done to achieve special effects such as a three-dimensional appearance. You can see some examples in Figure 5–2.

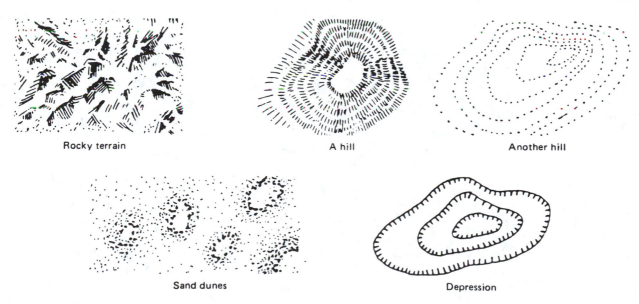

Rocky terrain · A hill · Another hill · Sand dunes · Depression

FIGURE 5–2. Graphic techniques for drawing topographical features.

FIGURE **5–3.**

You may even draw a representation of water features with rapids and whirlpools, depending on the purpose of your map (see Figure 5–3).

Some maps may show specific crops grown in an area, or information about how land is being used. Some optional symbols are shown in Figure 5–4.

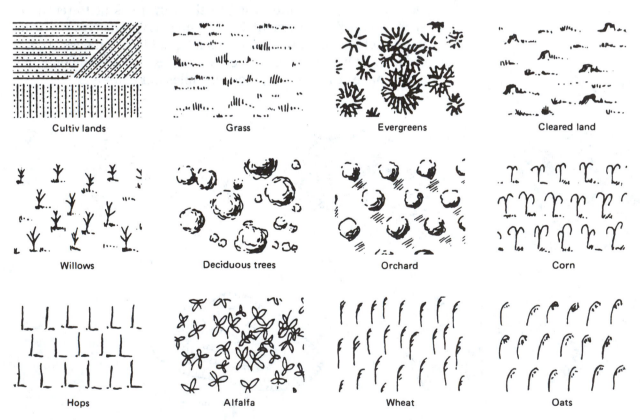

FIGURE **5–4.** Optional vegetation symbols.

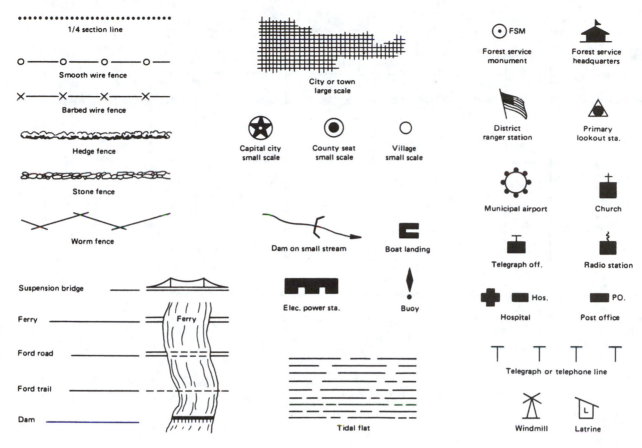

FIGURE 5-5. More possible map symbols.

There are probably as many symbols as there are purposes for maps. You have seen a lot of standard symbols; there are many more that you may find on a map, and there also may be some minor differences from one map preparer to the next. Figure 5–5 displays some additional map symbols.

DRAWING MAPPING SYMBOLS WITH CAD

Computer-aided drafting symbols are available for a variety of drafting applications. Most CAD programs have mapping symbols that are referred to as *symbol libraries, tablet menus,* and *template overlays*. A symbols library may be accessed through a digitizer tablet menu. A *digitizer tablet* contains a plastic or paper menu that displays commands and symbols used for drawing. The drafter uses a puck or stylus to pick the desired items from the menu. CAD programs that may be customized, such as AutoCAD, allow you to use symbol libraries designed by others, or you can create your own digitizer tablet symbol library overlay. Figure 5–6

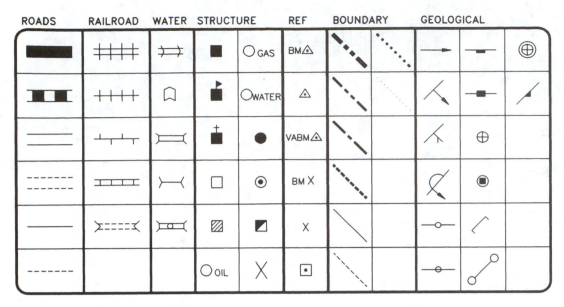

FIGURE 5–6. A customized tablet menu mapping symbol library for CAD.

shows a mapping template overlay that fits over the standard AutoCAD tablet menu. This allows you to use all of the AutoCAD commands as well as select mapping symbols for insertion in the drawing when needed.

Creating Your Own CAD Mapping Symbols

One of the greatest time-saving features of CAD is the ability to draw a symbol, save it, and then call it up for use any time the symbol is needed. CAD systems refer to these stored symbols as *blocks* or *cells*. AutoCAD, for example, uses the BLOCK and WBLOCK commands to create and store drawing symbols for later use. With this system a symbol can be inserted as many times as needed, and, when inserted, the symbol can be scaled and rotated to fit the drawing requirements.

Follow these guidelines when creating a symbol for future use:

- Make a sketch of the symbol.
- Establish an insertion point. This is a point or location on the symbol that targets where the symbol is placed on the drawing. Figure 5–7 shows common mapping symbols and their insertion points for placement on drawings.
- Give the symbol a name and record a copy of the symbol, its name, and the insertion point in a symbol catalog for future reference.

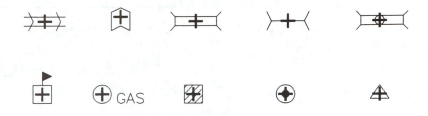

FIGURE 5–7. The insertion point for several mapping CAD symbols.

$+$ THIS SYMBOL REPRESENTS THE INSERTION POINT

When designing the symbol make it one unit wide by one unit high. This allows the symbol to be inserted in relationship to the drawing units, whether they are feet or meters. This also gives you the flexibility to easily scale the symbol to fit the requirements of any drawing when it is inserted. Figure 5–8 show the steps in creating a symbol for future use.

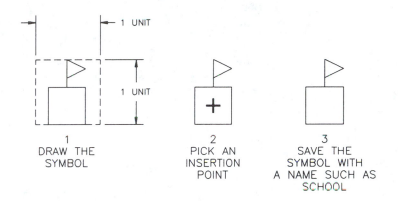

FIGURE 5–8. Creating a custom CAD mapping symbol.

$+$ THIS SYMBOL REPRESENTS THE INSERTION POINT

TEST

Part 1

Define each of the following types of map symbols. Use your best free-hand lettering.

5–1 Culture

5–2 Relief

5–3 Water

5–4 Vegetation

Part II

In the space provided, carefully sketch an example of each of the following map symbols.

5–1 Hard-surface heavy-duty road

5–2 Hard-surface medium-duty road

5–3 Railroad, single track

5–4 Bridge and road

5–5 Building (dwelling)

5–6 Power transmission line

5–7 Perennial stream

5–8 Intermittent stream

5–9 Marsh

5–10 Orchard

Part III

Using shading techniques, sketch examples of each of the following features.

5–1 Rocky terrain

5–2 Hill

5–3 Depression

5–4 Sand dune

5–5 River

PROBLEM

P5–1 Draw a map in the space provided on Figure P5–1. Use the following instructions to draw the map:

1. Use standard topographic symbols.
2. Use normal drafting practices with pencil, ink or CAD as directed by your instructor.
3. Use appropriate mapping colors unless otherwise instructed.
4. North is toward the top of the page.
5. The scale is 1 in. = 200 ft.
6. From point *A*, draw a hard-surface medium-duty road with a bearing of N60°E.
7. Beginning 975 ft from point *A*, draw a bridge over a 40-ft-wide river.
8. The center of the river runs through points *B*, *C*, *D*, and *E*. Use freehand lines to draw the river unless CAD is used.
9. At 420 ft from point *A* along the hard-surface road, draw an improved light-duty road with a bearing of S26°E.
10. A small creek enters the river at point *C* with a bearing of N45°E for 385 ft (tangent point), then turns with a radius of 200 ft and a bearing of N68°W.
11. North of the medium-duty road and southwest of the river, the entire area is orchard to within 200 ft of the river.
12. The area south of the medium-duty road and west of the light-duty road is a corn crop.
13. On both sides of the small creek is a grove of deciduous trees 100 ft wide.
14. All other areas are grasslands.
15. Two hundred feet northeast of the bridge and on the north side of the medium-duty road is a forest service headquarters.

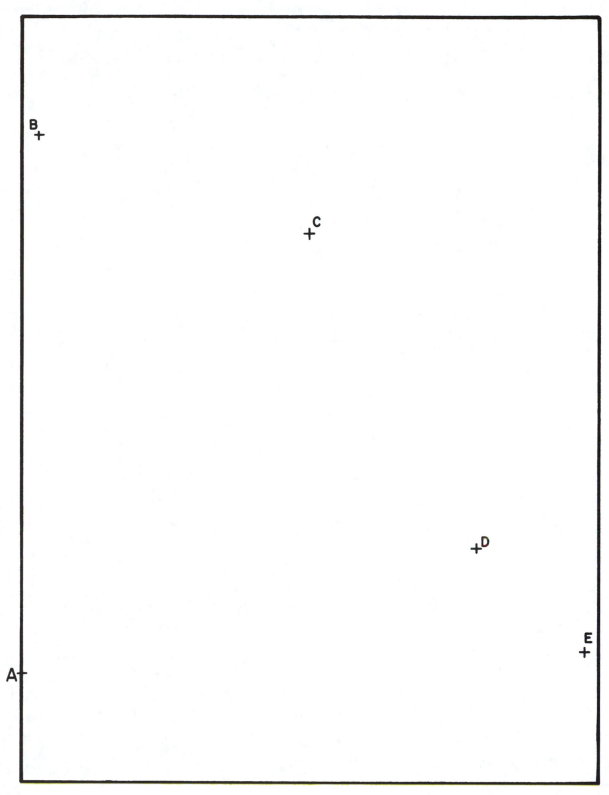

FIGURE **P5–1.**

CHAPTER 6

Legal Descriptions and Plot Plans

This chapter shows the three typical ways of identifying property. Each individual property must be completely described in a survey. This survey becomes the legal description that keeps your land or property separate from your neighbor's property.

The topics covered include:

1. Metes and bounds
2. Lot and block
3. Public land surveys
4. Basic reference lines
5. Rectangular system
6. Plot plans
7. Methods of sewage disposal

METES AND BOUNDS

Metes and bounds is a method of describing and locating property by measurements from a known starting point called a *monument*. Metes can be defined as being measurements of property lines expressed in units of feet, yards, rods or meters. Bounds are boundaries such as streams, streets, roads, or adjoining properties. The monument, or *point of beginning* of the system, is a fixed point such as a section corner, a rock, tree, or the intersection of streets.

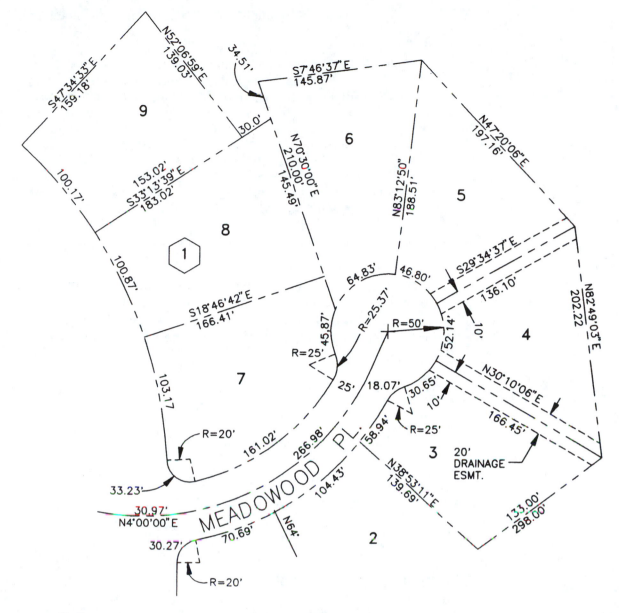

FIGURE 6-1. A typical plat using metes and bounds.

The metes and bounds system is often used for describing irregularly shaped plats and, while used in all areas, it is used as the primary method of describing plats in states east of the Mississippi River. A typical plat using metes and bounds is shown in Figure 6–1.

The following is a sample legal description using metes and bounds:

> BEGINNING at the intersection of the centerline of W. Powell Boulevard, formerly Powell Valley Road and the centerline of S.W. Cathey Road; thence running East along the centerline of W. Powell Boulevard 184 feet; thence South on a line parallel with S.W. Cathey Road, 200 feet; thence West on a line parallel with W. Powell Boulevard, 184 feet; thence North along the centerline of S.W. Cathey Road to the place of beginning; EXCEPTING therefrom, however, the rights of the public in and to that portion of the herein described property lying within the limits of W. Powell Boulevard and S. Cathey Road.

LOT AND BLOCK

Lot and block is a method that describes land by referring to a recorded plat, the lot number, the county, and state. A legal lot and block must be filed with the county clerk or recorder as part of a plat, which is a map or plan of a subdivision. Look at Figure 6–2.

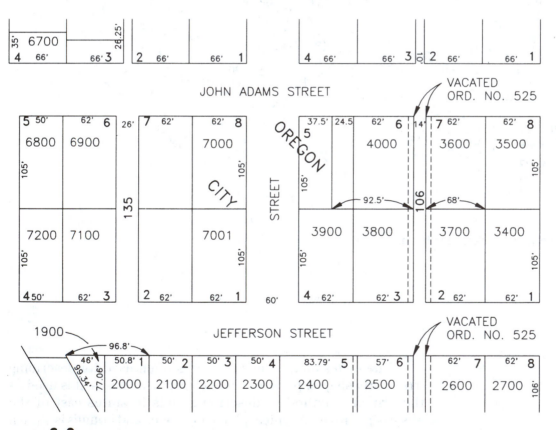

FIGURE 6–2. Lot and block.

The lot and block system is commonly used to describe small units of property in a subdivision. The exact boundaries of the subdivision may be described by the rectangular system or the metes and bounds system. For example, the property lines of the lots in the small subdivision shown in Figure 6–1 are established using metes and bounds; but the lots can additionally be identified with a lot and block system. Just remember that a necessary part of a plot plan, or plat, is the inclusion of an accurate legal description.

The following is an example of a lot and block description:

Lot 7, Block 135, Oregon City Subdivision, City of Oregon City, Clackamas County, State of Oregon.

RECTANGULAR SYSTEM

Public Land Surveys

In the midwestern and far western states, or *public land states*, the U.S. Bureau of Land Management devised a *rectangular system* for describing land. The states involved in the public land surveys are: Alabama, Alaska, Arizona, Arkansas, California, Colorado, Florida, Idaho, Illinois, Indiana, Iowa, Kansas, Louisiana, Michigan, Minnesota, Mississippi, Missouri, Montana, Nebraska, Nevada, New Mexico, North Dakota, Ohio, Oklahoma, Oregon, South Dakota, Utah, Washington, Wisconsin, and Wyoming. In 1850, the federal government bought 75 million acres from Texas, and these too are public lands. In the list above, you will note some southern, and even one southeastern state (Florida), included in the public land states. The public land states begin with Ohio. Its west boundary is the first principal meridian.

Basic Reference Lines

Each large portion of the public domain is a single *great survey*, and it takes on as much as it can use of one parallel of latitude and one meridian of longitude. The initial point of each great survey is where these two basic reference lines cross. This must be determined astronomically: a star-true point. The parallel is called the *base line*, and the meridian is called the *principal meridian*. There are 31 pairs or sets of these standard lines in the United States proper, and three in Alaska.

At the outset, each principal meridian was numbered. However, the numbering stopped with the sixth principal meridian, which passes Nebraska, Kansas, and Oklahoma. The balance of the 31 sets of standard lines took on local names. For example, in Oregon the public land surveys use the *Willamette meridian* for

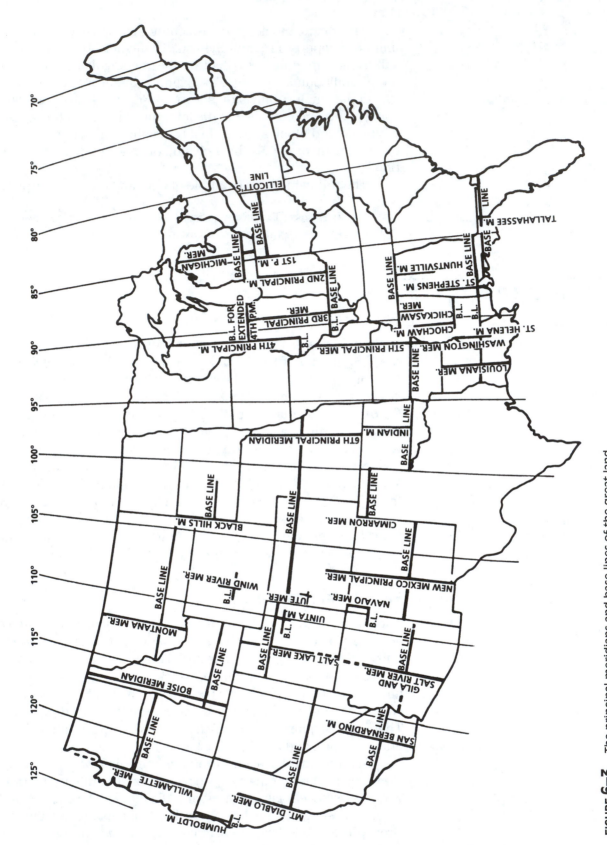

FIGURE 6-3. The principal meridians and base lines of the great land survey, not including Alaska.

the principal meridian. The principal meridians and base lines of the great land survey, not including Alaska, are shown in Figure 6–3.

Rectangular-Township/Section System

Using base lines and meridians, an arrangement of rows of blocks, called *townships*, is formed. Each township is 6 miles square. They are numbered by rows or tiers, the rows running east–west. These tiers are counted north and south from the base line. But instead of saying "tier one, tier two, tier three, . . . ," we say "township." For example, a township in the third tier north of the base line is named, "Township No. 3 North," abbreviated as "T.3N." Similarly, the third tier south of the base line is named "T.3S."

Townships are also numbered according to which vertical (north–south) column they are in. These vertical columns of townships are called *ranges*. Ranges take their numbers east or west of the principal meridian. A township in the second range east of a principal meridian is "R.2E."

If you put these two devices of location together, the township in the third tier north of the base line and in the second range east of the principal meridian is "T.3N.,R.2E." Figure 6–4 illustrates the arrangement of townships about the two reference lines. The tiers number as far north and south, and ranges number as far east and west, as that particular great public land survey goes.

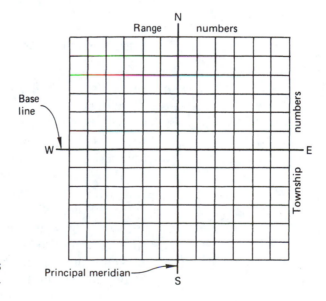

FIGURE 6–4. Arrangement of townships about the base line and principal meridian.

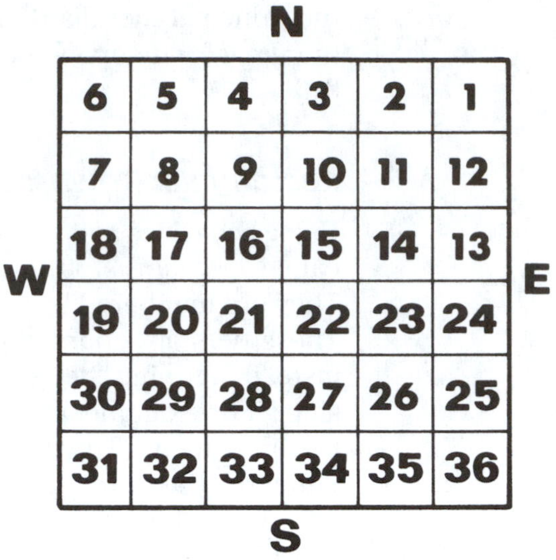

FIGURE 6–5. Correct method of numbering sections within a township.

The Division of Townships

Figure 6–5 illustrates the division of a township into sections, and how the sections are numbered in each township.

A township is a square with sides approximately 6 mi long. At each mile along the 6-mi sides of a township, a line cuts across, forming a checkerboard of 36 squares, each a mile square. Each parcel of land, being approximately 1 mi square, contains 640 acres. These squares are *sections*. Sections are numbered 1 to 36, beginning at the northeast corner of the township and going across from right to left, then left to right, right to left, and so on, until all 36 squares or sections are numbered. Remember, always start numbering with 1 in the upper right-hand corner, and 36 will always be in the lower right corner.

The Subdivision of Sections

Figure 6–6 illustrates some typical subdivisions of a section. The smallest subdivision shown in Figure 6–6 is 10 acres. However, even smaller subdivisions can be made, such as a residential subdivision of lots less than an acre in size.

Each section, as shown in Figure 6–6, may be subdivided into *quarter sections*. These are sometimes called *corners*: for example, "northwest corner of Section 16" or "NW 1/4 of Sec. 16."

A quarter section may be further divided into *quarter-quarters*; for example, "NW 1/4, NW 1/4 of Sec. 36." Further subdivisions of a quarter-quarter are illustrated in Figure 6–6.

In a written description of a portion of land, it will be noted that the smallest portion is written first, followed by the next larger portion, and so on. Looking at Figure 6–6 again, the smallest portions are two 10-acre subdivisions. Note that the description begins with the smallest division and progresses to the largest: for

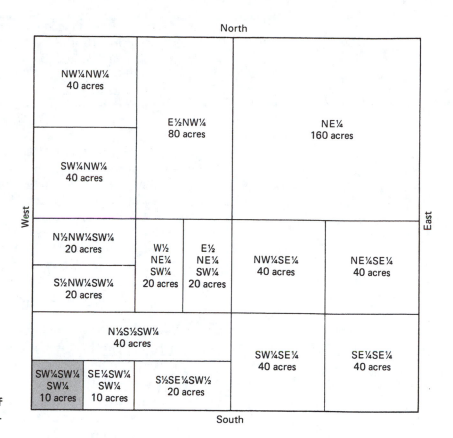

FIGURE 6–6. Some typical subdivisions of a section.

example, "SE 1/4, SW 1/4, SW 1/4" as represented by the shaded area in Figure 6–6.

The Complete Description

The description of a small subdivision must include its relationship with the township, and then in turn relate to the reference lines of the great survey. Always remember that a complete description begins with the smallest division and progresses to the largest. An example of a legal description could read like this:

> The SW 1/4, SE 1/4 of Section 9 in Township 3 South, Range 2 East of the Willamette meridian, all being in Clackamas County, Oregon.

The following legal description reveals that a complete description of real property may include all three types of descriptions in combination. This example uses the rectangular system to identify the point of beginning, metes and bounds to describe the boundary lines, and a lot and block description that may also be used as an alternate description.

> Part of the Stephen Walker D.L.C. No. 52 in Sec. 13 T. 2 S. R. 1E., of the W.M., in the County of Clackamas and State of Oregon, described as follows:

Beginning at the one-quarter corner on the North line of said Section 13; thence East 111.65 feet; thence South 659.17 feet; thence South 25°35′ East 195.00 feet to a ⅝-inch iron rod; thence South 56° West 110.0 feet to a ⅝-inch iron rod; thence South 34° East 120.79 feet to a ⅝-inch iron rod; thence South 19°30′ East 50.42 feet to a ⅝-inch iron rod; thence South 40°32′ East 169.38 feet to a ⅝-inch iron rod; thence South 49°29′ West 100.0 feet to a ⅝-inch iron rod; thence South 65°00′50″ West 51.89 feet to a ⅝-inch iron rod; thence South 52°21′30″ West 125.0 feet to a 1/2-inch iron pipe, being the true point of beginning; thence continuing South 52°21′30″ West 124.91 feet to a ½-inch iron pipe on the arc of a circle with a radius of 50.0 feet, the center point of which bears North 22°01′30″ West 50.0 feet from said last-mentioned iron pipe; thence Northeasterly along the arc of said circle, through an angle of 69°53′29″, 60.99 feet to a ½-inch iron pipe; thence North 88°05′01″ East 51.00 feet to a ½-inch iron pipe; thence South 39°55′35″ East 80.0 feet to the true point of beginning. ALSO known as *Lot 6, Block 3, CHATEAU RIVIERE*, in Clackamas County, Oregon.

Figure 6–7 displays a small subdivision drawn with a CAD system. The lot lines are shown with lengths and bearings. The survey can be combined with aerial photography to produce CAD drawings that have added realism. When a subdivision is first drawn with the CAD system, true north for the preliminary drawing is shown pointing to the top of the screen. This allows the greatest degree of accuracy when entering lengths and bearings, and also allows map information from other sources (such as aerial photography which also uses north pointing to the top of the screen) to be imported exactly.

When the initial drawing with all its components is completed, it may be necessary to rotate the entire drawing to more easily fit on the plotted page. It is important at this point to maintain

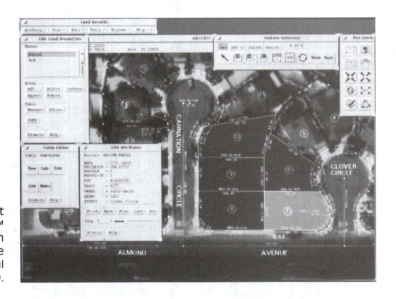

FIGURE 6–7. Land records management is a common use of GIS and ARC COGO™ tools. This image was created with ARC/INFO® software (Graphic image supplied courtesy of Environmental Systems Research Institute, Inc.).

the precision of the original survey(s). One way that this can be done is to rotate the entire drawing from an origin of 0,0 by the appropriate number of degrees. It is helpful to make a note in the margin of your drawing stating the number of degrees the drawing was rotated, remembering that no lines that you now obtain information about on the drawing will give true bearings.

PLOT PLANS

A *plat*, or plot plan, is a map of a piece of land. A plat becomes a legal document and contains an accurate drawing, as well as a written description of the land. It is not always necessary to show relief as contour lines. Often, only arrows are used to show direction of slope. In some specific situations, however, actual contour lines may have to be established and drawn. This may be necessary when the land has unusual contours, is especially steep, or has out-of-the-ordinary drainage patterns. Be sure to check with your jurisdiction for local requirements.

Requirements

Many items are necessary to make a plot plan a legal, working document. As a drafter, you can use the data as a checklist for proper completion of your plot. Be sure to check your local city or county regulations for any different requirements. Some of the following requirements may not apply to your plot plan. Many building departments require plot plans to be drawn on 8½ by 14-in. paper.

1. Legal description of the property.
2. Property lines, dimensions, and bearings.
3. Direction of north.
4. All roads, existing and proposed.
5. Driveways, patio slabs, parking areas, and walkways.
6. Proposed and existing structures.
7. Location of well and/or water service line. Location of wells on adjacent properties.
8. Location of proposed gas and power lines.
9. Location of septic tank, drainfield, drainfield replacement area, and/or sewer lines.
10. Dimensions and spacing of soil absorption field, or leach lines, if used.
11. Location of soil test holes, if used.

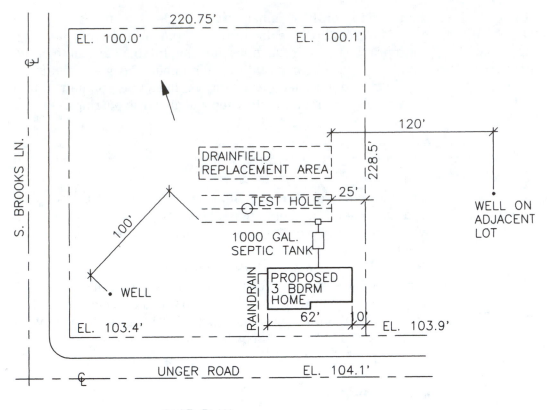

220.75'

EL. 100.0'　　　　　　　　　　　EL. 100.1'

S. BROOKS LN.

120'

228.5'

DRAINFIELD
REPLACEMENT AREA

TEST HOLE　　　25'

1000 GAL.
SEPTIC TANK

WELL ON
ADJACENT
LOT

100'

WELL

RAINDRAIN

PROPOSED
3 BDRM
HOME

62'　　10'

EL. 103.4'　　　　　　　　　　EL. 103.9'

UNGER ROAD　　　EL. 104.1'

PLOT PLAN
SCALE: 1"=50'

SECTION 17 T.5S., R.8E. WM
CLACKAMAS COUNTY, OREGON
TAX LOT NO. 1876

FIGURE 6–8. A typical plot plan.

12. Proposed location of rain drains, footing drains, and method of disposal.

13. Ground elevation at lot corners, and street elevation at driveway centerline.

14. Slope of ground.

15. Proposed elevations of main floor, garage floor, and basement or crawl space.

16. Number of bedrooms proposed.

17. Proposed setback from all property lines.

18. Utility and drainage easements.

19. Natural drainage channels.

20. Total acreage.

21. Drawing scale: for example, 1 in. = 50 ft. (1″ = 50′)

Figure 6–8 shows a typical plot plan. Notice that not all of the information is identified. Be sure to use all of the information that you need to describe your plot completely.

Septic Tank

The conventional septic tank is usually a concrete or steel box where the wastewater from the house collects. Wastewater from toilets, bathtubs, showers, laundry, and kitchen is fed into this tank. It is designed to hold the water for two or three days, long enough for most of the heavy suspended material to sink to the bottom of the tank to form a sludge. Lighter, floating materials float to the top of the tank, where they remain trapped between the inlet and outlet pipes. After a couple of days, the wastewater portion leaves the tank as *effluent*. The effluent is discharged to the underground piping network, called a *soil absorption field*, *drainfield*, or *leach lines*.

The soil absorption field is an underground piping network buried in great trenches usually less than 2 ft below the surface of the ground. This field distributes the effluent over a large soil area, allowing it to percolate through the soil. The soil usually acts as an excellent filter and disinfectant by removing most of the pollutants and disease-causing viruses and bacteria found in the effluent. Figure 6–9 shows a section through a septic tank and a partial absorption field.

Now take a look at Figure 6–10 and you will see how the septic system should be drawn on a plot plan. Keep in mind that specific lengths of drainfield and minimum specifications are determined by local requirements. Be sure that the drainfield lines run parallel to the contour lines.

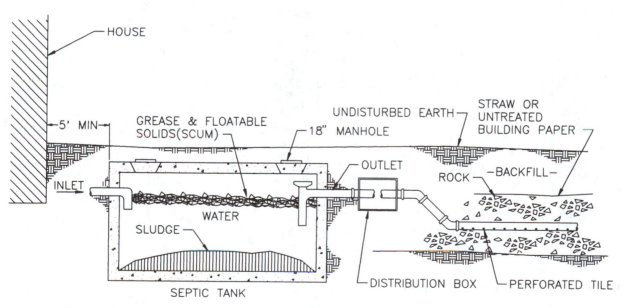

FIGURE 6–9. Cross section through a typical septic tank.

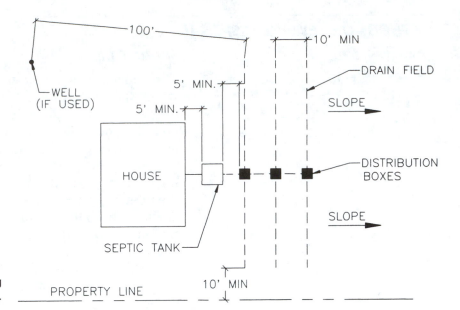

FIGURE 6–10. Sample plot plan showing a house and septic system.

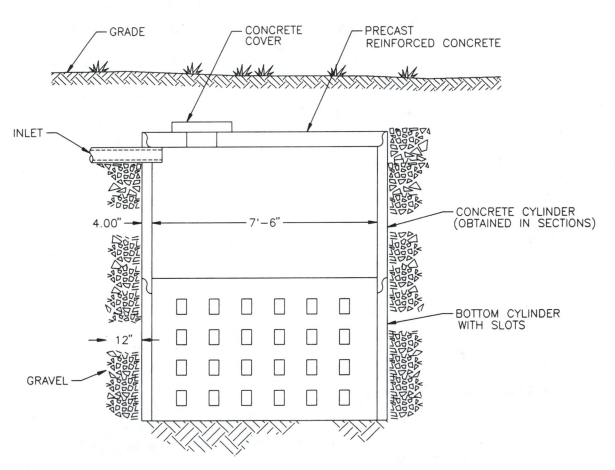

FIGURE 6–11. Cross section of a typical cesspool.

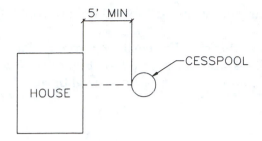

FIGURE 6–12. Sample plot plan showing a house and cesspool.

Cesspools

Cesspools have the same purpose as septic systems, that is, the breaking down and distribution of waste materials to an area of earth. The soil then acts as a filter to disperse pollutants. Cesspools are used in locations where the soil bearing strata is very porous. This would be an area of gravel or similar material of considerable depth. The structure that makes up the cesspool is a large concrete cylinder. This cylinder can be of precast concrete, concrete block, or other materials. The cesspool has slots at the bottom for the effluent to escape into a layer of gravel around the tank and then into the soil, made up of porous material. Your local soils department will be able to advise you as to the type of system that should be used. A cross section of a typical cesspool is shown in Figure 6–11.

Now take a look at Figure 6–12 and you can see how a cesspool will look on a plot plan.

Public Sewers

In locations where public sewers are available, the plot plan should show a sewer line from the house to the public sewer, usually located in the street or in an easement provided somewhere near the property. This method of sewage disposal is often easier than the construction of a cesspool or septic system. See Figure 6–13 for a cross section of a sewer hookup.

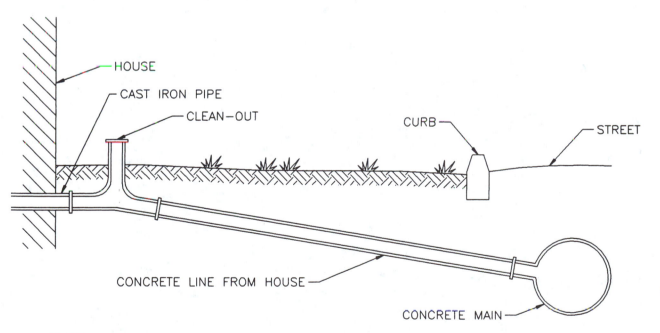

FIGURE 6–13. Cross section of a typical public sewer installation.

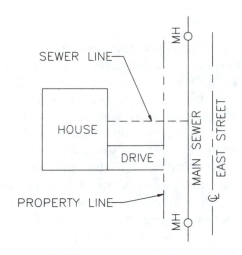

FIGURE 6–14. Sample plot plan showing a house and public sewer layout.

Figure 6–14 shows how a conventional sewer hookup would be drawn on a plot plan.

A sewer hookup such as this could be easily incorporated in a CAD drawing like the one shown in Figure 6–7. Figure 6–15, drawn with a CAD system, shows a portion of a neighborhood with a public sewer layout.

Utilities

The other utilities may be drawn into the property from the main lines that exist in a street or utility easement. These utilities may include electrical, gas, phone, and TV cable. Some utilities may be overhead, such as electrical. However, all utilities could be brought into the property from underground. Before you complete the plot plan, determine where the utilities will enter the property, so that they can be identified on the proposed plan.

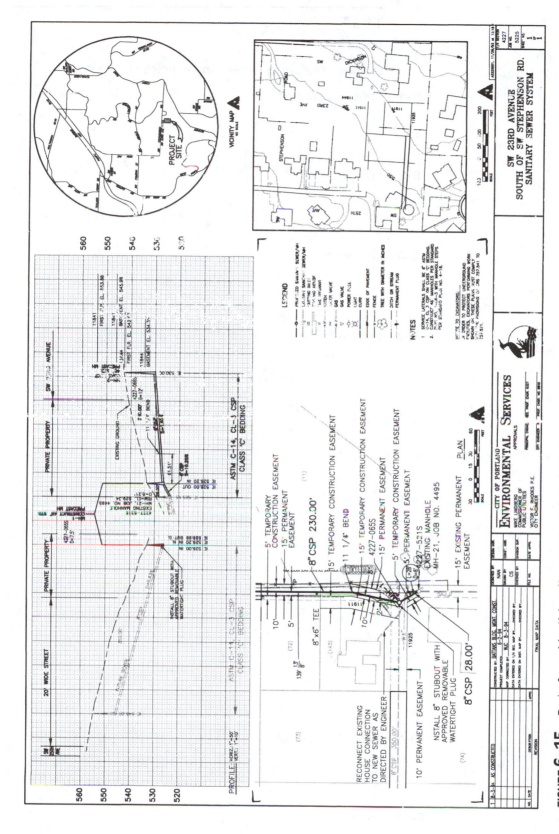

FIGURE 6–15. Part of a residential sanitary sewer system drawing. (Courtesy of The City of Portland, Portland, OR.)

TEST

Part I

Define the following terms using concise statements.

6–1 Metes and bounds

6–2 Lot and block

6–3 Township

6–4 Section

6–5 Plot plan

6–6 Septic tank

6–7 Cesspool

6–8 Base line

6–9 Principal meridian

6–10 Acre

Part II

Given the section shown in Figure T6–1 with areas labeled by letters, provide the legal description and the number of acres for each area.

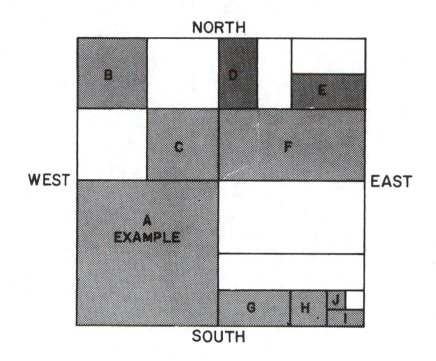

	Area	Legal Description	Acres
6.1	A	SW 1/4	160
6.2	B		
6.3	C		
6.4	D		
6.5	E		
6.6	F		
6.7	G		
6.8	H		
6.9	I		
6.10	J		

FIGURE **T6–1**

Part III

Carefully sketch examples of plot plans that will display each of the following characteristics.

6–1 Public sewer to house on plot. Show driveway to house from street.

6–2 Cesspool to house on plot. Show driveway to house from street.

6–3 Septic system to house on plot. Show driveway to house from street.

PROBLEMS

Part I: Drawing Plot Plans from Sketches

Use one of the following drafting methods and media as specified by your instructor:

Use pencil, with freehand lettering and proper line technique on vellum.

Use ink and mechanical lettering equipment on vellum.

Use ink and mechanical lettering equipment on polyester film.

Use computer-aided drafting and make a print or plot of your drawing.

P6–1   Given the rough sketch shown in Figure P6–1, draw a plot plan on 8½-by 14-in. vellum. The sketch is not to scale. Scale your drawing to fit the paper with a ½-in. minimum margin to the edge of the paper. Use the following information:

1. You select the scale (e.g., 1 in. = 20 ft, 1 in. = 50 ft). The scale you select should make the plot plan as large as possible within the limits of the paper.
2. Label the plot plan, scale, legal description, elevations given, and north arrow.
3. Use as many of the plot plan requirements described in this chapter as possible.
4. Property lines run north/south and east/west.

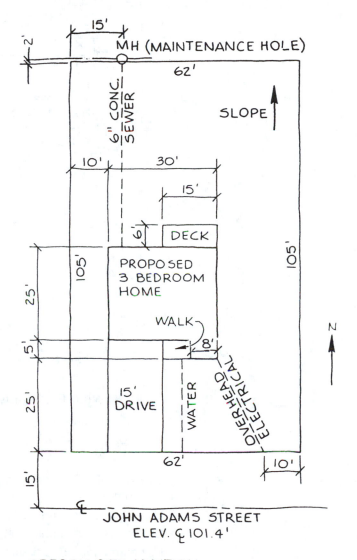

OREGON CITY ADDITION
LOT 7, BLOCK 2, OREGON CITY, OREGON
ELEV. MAIN FL. 101.8'
FIGURE **P6–1**   ELEV. BSMT. FL. 92.8'

P6–2 Given the rough sketch in Figure P6–2, draw a plot plan on 18- by 24-in. or 17- by 22-in. media. The sketch is not to scale. Scale your drawing for best utilization of the paper size. Use the following information:

1. Construct your own title block with title, legal description, your name, the scale, and a north arrow.
2. Use as many of the plot plan requirements described in this chapter as possible.

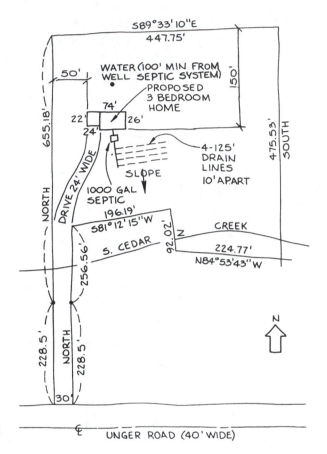

FIGURE P6–2

Part II: Drawing Plot Plans from Written Information

Draw the following property boundaries from the given legal descriptions or property line data. Use one of the following drafting methods and media as specified by your instructor:

Use pencil, with freehand lettering and proper line technique on vellum.

Use ink and mechanical lettering equipment on vellum.

Use ink and mechanical lettering equipment on polyester film.

Use computer-aided drafting and make a print or plot of your drawing.

Use the appropriate scale and sheet size unless otherwise specified. Give the general legal description, scale, a north symbol, and other information requested with each problem.

P6–3 All that portion of Lot 140, commencing at the southwesterly corner and thence continuing along said parallel line, North 0 Degrees 22 Minutes East, 50.00 feet; thence South 89 Degrees 38 Minutes East, 270.85 feet to a point in the center line of Hillside Drive; thence South 58 Degrees 28 Minutes West, 12.98 feet to an angle point in said center line; thence continuing along said center line, South 36 Degrees 42 Minutes West, 102.0 feet; thence leaving said center line, North 78 Degrees 34 Minutes West, 203.46 feet to the TRUE POINT OF BEGINNING.

P6–4 SITE DESCRIPTION

Starting at a point on the west side of the 35′ right of way of El Corto Street, San Marcos California:

Station	Bearing	Distance
1 P.O.B.	S 75 00 E	310.01′
2	N 23 00 E	276.06′
3	N 27 30 W	221.63′
4	S 23 30 W	420.07′
5	N 75 00 W	143.32′
6	S 02 45 W	22.23′

Legal Description:

A portion of Lot #9, in block "L" of Charles Victor Hall Tract Unit #2, Map No. 2056 filed in the office of the County Recorder of San Diego County, State of California, on September 22, 1927.

P6–5 That portion of Lot 6, commencing at the most Northerly corner of that parcel of land delineated and designated as "1.17 Acres (Net)": thence South 55 Degrees 17 Minutes 37 Seconds East, 203.85 feet to the Northwesterly boundary of County Road (Highland Drive) being a point on the arc of a 165.88 foot radius curve concave Northwesterly, a radial line of said curve bears South 13 Degrees 38 Minutes 13 Seconds East to said point; thence along said Northwesterly boundary as follows: Westerly along the arc of said curve through a central angle of 10 Degrees 17 Minutes 13 Seconds a distance of 29.78 feet to the beginning of a reverse 170.0 foot radius curve; Southwesterly along the arc of said curve through a central angle of 52 Degrees 40 Minutes a distance of 156.27 feet to the point of tangency in the Southeasterly line of said Lot 6; and along said Southeasterly line tangent to said curve South 33 Degrees 59 Minutes West, 37.91 feet to the most Southerly corner of said land; thence north 56 Degrees 01 Minute West, 115.00 feet to a line which bears South 33 Degrees 59 Minutes West from the True Point of Beginning; thence North 33 Degrees 59 Minutes East, 195.73 feet to the TRUE POINT OF BEGINNING.

Part III: Converting Plat Map Drawings to Formal Drawings

Draw the following formal plat drawings from the given plat maps. The plat maps are not to scale, so you will have to select a scale such as 1 in. = 10 ft or 1 in = 50 ft, for example. Select the paper size to fit each problem. Use one of the following drafting methods and media as specified by your instructor:

Use pencil, with freehand lettering and proper line technique on vellum.

Use ink and mechanical lettering equipment on vellum.

Use ink and mechanical lettering equipment on polyester film.

Use computer-aided drafting and make a print or plot of your drawing.

Use the appropriate scale and sheet size unless otherwise specified. Give the correct line representation, all curve data, general legal description, scale, a north symbol, and other information requested with each problem. Make a print for instructor evaluation.

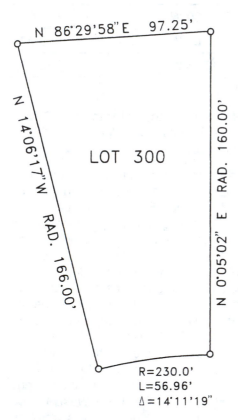

FIGURE **P6-6**

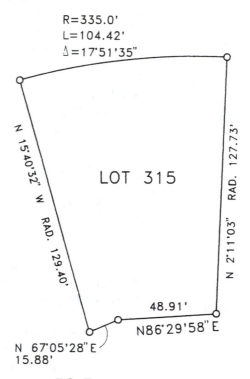

FIGURE **P6-7**

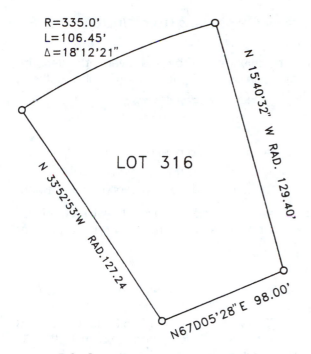

R=335.0'
L=106.45'
Δ=18°12'21"

LOT 316

N 15°40'32" W RAD. 129.40'

N 33°52'53"W RAD.127.24

N67D05'28" E 98.00'

FIGURE **P6–8**

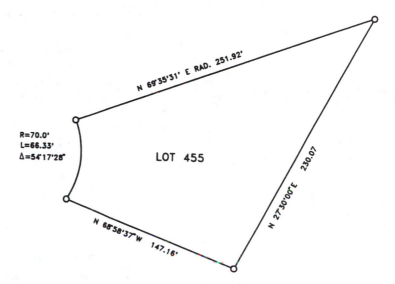

N 69°35'31' E RAD. 251.92'

R=70.0'
L=66.33'
Δ=54°17'28"

LOT 455

N 27°30'00"E 230.07

N 68°58'37" W 147.16'

FIGURE **P6–9**

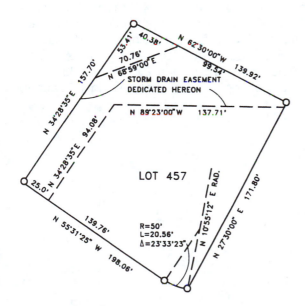

53.41' 40.38'
N 62°30'00" W
70.76'
157.70' N 68°59'00" E 99.54' 139.92'
N 34°28'35" E
STORM DRAIN EASEMENT
DEDICATED HEREON
N 89°23'00" W 137.71'
94.08'
N 34°28'35" E
LOT 457
N 10°55'12" E RAD.
N 27°30'00" E 171.80'
25.0' N 34°28'35" E
N 55°31'25" W
139.76'
R=50'
L=20.56'
Δ=23°33'23"
198.06'

FIGURE **P6–10**

Part IV: Advanced Drawing Plot Plans from Written Information

Draw the following property boundaries from the given legal descriptions or property line data. Use one of the following drafting methods and media as specified by your instructor:

Use pencil, with freehand lettering and proper line technique on vellum.

Use ink and mechanical lettering equipment on vellum.

Use ink and mechanical lettering equipment on polyester film.

Use computer-aided drafting and make a print or plot of your drawing.

Use the appropriate scale and sheet size unless otherwise specified. Give the general legal description, scale, a north symbol, and other information requested with each problem.

P6–11 LEGAL DESCRIPTION

Starting at a point on the north side of the 66 feet right of way of Edgewood Drive; thence N 61° 56' 33" E for a length of 184.24 feet; thence N 1° 16' 55" W for a length of 715.37 feet; thence S 89° 44' 40" W for a length of 164.51 feet; thence South 1° 16' 55" E for a length of 801.32 feet to the point of beginning.

P6–12

All that portion of Lot Four (4) in Block One Hundred Forty-nine (149) of the City of Escondido, as per Map thereof No. 336, filed in the office of the County Recorder of said San Diego County, July 10, 1886, described as follows:

Beginning at a point on the center line of Grant Avenue, North 69° 43' East, a distance 240.0 feet from the center line intersection of North Broadway and Grant Avenue, which said letter point is the Northwest corner of said Lot Four (4) of Block One Hundred Forty-nine (149); thence North 69° 43' East along the center line of said Grant Avenue, a distance of 317.0 feet to a point; thence South 00° 57' East 96.0 feet; thence South 10° 40' W. 166.0 feet; thence South 69° 43' W. 200.0 feet; thence North 20° 17' W. 233.0 feet; more or less, to the point of beginning, and containing 1.42 acres, more or less.

LEGAL PROPERTY DESCRIPTION: PLOT AT SCALE 1″ = 100.0′
P.O.B. is the center of Bear Road which is 66.0 feet wide.

Bearing	Distance
N 3 45 E	560.5
N 0 15 W	619.2
N 45 00 E	425.3 Top of river bank
N 45 00 E	80.6 Center of Indian River
S 78 10 E	232.0
S 59 45 E	238.5
S 50 45 E	270.7
S 23 15 E	318.8
S 71 30 E	200.4
S 7 45 W	430.5 Top of river bank
S 7 45 W	114.5 Center of Indian River
DUE WEST	451.3
S 0 10 W	397.3
N 79 00 W	465.1
N 88 47 W	365.0 To P.O.B.

Acreage Computation:

1 acre = 43,560 SF

P6–14

Property Legal Description
Flemming Property
Corner of Birmingham and MacKinnon
Cardiff, California
File No. SB5501

Point No.	Bearing		Distance
47	N 0° 52′ 0″ E		95.0
39	N 27° 28′ 50″ E		19.875
42	N 25° 36′ 2″ E		102.60
165	ARC CENTER		
	R = 102.60	DELTA = 12 39 1	L = 22.556
	S 38° 15′ 3″ W	102.60	
112	S 38° 15′ 3″ W		130.00
162	ARC CENTER		
	R = 130.0	DELTA = 23 26 9	L = 53.174
	N 14° 48′ 54″ E	130.0	
116	S 63° 53′ 58″ W		25.04
117	S 73° 22′ 11″ W		23.25
118	S 77° 50′ 6″ W		23.51
119	S 68° 44′ 46″ W		23.41
120	S 1° 24′ 39″ W		52.23
124	N 89° 53′ 24″ E		99.00
129	S 2° 5′ 19″ W		64.988
45	N 89° 56′ 50″ E		50.00
47			

Plot above description in the best scale, and compute the total area in square feet and in acres.

P6–15

All that portion of Lot 12 of MARTIN'S ADDITION TO VISTA, in the County of San Diego, State of California, as shown on Map No. 1472, on file in the Office of the County Recorder of said San Diego County, described as follows:

Beginning at a point of intersection of the Southeasterly line of said Lot 12, with a line drawn parallel with and distant 140.00 feet, measured at right angles Northeasterly from the Northeasterly line of Citrus Avenue, 60 feet wide, as shown on said Map No. 1472, said point being the Southeasterly corner of the land conveyed to William Allington by deed dated February 02, 1933, and recorded in Book 188 Page 391 of Official Records of said San Diego County; thence North 30 Degrees 18 Minutes West along the Easterly line of land so conveyed, being also the aforementioned parallel line, 186.45 feet to an angle point in said Easterly line; thence continuing along said Easterly line, being parallel with and distant 140.00 feet at right angles Easterly from the Easterly line of said Citrus Avenue, North 0 Degrees 22 Minutes East 141.97 feet to the TRUE POINT OF BEGINNING; thence continuing along said parallel line, North 0 Degrees 22 Minutes East, 50.00 feet to a point distant thereon South 0 Degrees 22 Minutes West, 100.00 feet from the Southerly line of the Northerly 4.0 Acres of said Lot 12, as said Northerly 4.0 Acres were conveyed to H. B. Morris, et ux, by deed dated April 14, 1930 and recorded in Book 1784 Page 187 of Deeds; thence parallel with and 100.00 feet Southerly at right angles from said Southerly line of said Northerly 4.0 Acres as conveyed to said Morris, South 89 Degrees 38 Minutes East, 270.85 feet to a point in the center line of Hillside Drive as shown on said Map No. 1472; thence along said center line of said Hillside Drive, South 58 Degrees 28 Minutes West, 12.98 feet to an angle point in said center line; thence continuing along said center line, South 36 Degrees 42 Minutes West, 102.0 feet; thence leaving said center line, North 78 Degrees 34 Minutes West, 203.46 feet to the TRUE POINT OF BEGINNING.

P6–16

That portion of Lot 6 in Block 4 of Keeney's Marine View Gardens, in the County of San Diego, State of California, according to Map thereof No. 1774, filed in the office of the County Recorder of San Diego County, December 31, 1923, lying within the following described boundary:

Commencing at the most Northerly corner of that parcel of land delineated and designated as "1.17 Acres (Net)" on Record of Survey Map No. 4270, filed in the office of the County Recorder of San Diego County, April 18, 1957, being also the most Northerly corner of land described in deed to John R. Minton, et ux, recorded December 23, 1964 as File No. 232023; thence along the Northeasterly line of said Minton's land, South 55° 17' 37" East, (Record South 55° 52' 00" East) 115.01 feet to the TRUE POINT OF BEGINNING; thence continuing along said Northeasterly line South 55° 17' 37" East, 203.85 feet to the Northwesterly boundary of County Road Survey No. 821 (Highland Drive) as described in Parcel 2 in deed to the County of San Diego, recorded April 21, 1949 as Document No. 35757 in Book 3179, Page 154 of Official Records; being a point on the arc of a 165.88 foot radius curve concave Northwesterly, a radial line of said curve bears South 13° 38' 13" East to said point; thence along said Northwesterly boundary as follows: Westerly along the arc of said curve through a central angle of 10° 17' 13" a distance of 29.78 feet to the beginning of a reverse 170.00 foot radius curve; Southwesterly along the arc of said curve through a central angle of 52° 40' 00" a distance of

156.27 feet to the point of tangency in the Southeasterly line of said Lot 6; and along said Southeasterly line tangent to said curve South 33° 59′ 00″ West, 37.91 feet to the most Southerly corner of said Minton's land; thence along the Southwesterly line of said Minton's land North 56° 01′ 00″ West, 115.00 feet to a line which bears South 33° 59′ 00″ West from the True Point of Beginning; thence North 33° 59′ 00″ East, 195.73 feet to the TRUE POINT OF BEGINNING.

ALSO that portion of Lot 6 adjoining the above described land as shown on said Record of Survey Map No. 4270 which lies Westerly of the Southeasterly prolongation of the Northeasterly line of the above described land.

CHAPTER 7

Contour Lines

The topography of a region is best represented by contour lines. The word topography comes from the Greek words *topos*, a place, and *graphein*, to draw. The most common method of "drawing a place" for mapping purposes is to represent differences in elevation with contour lines. These lines connect points of equal elevation. They can also reveal the general lay of the land and describe certain geographical features to those trained in topographical interpretation.

This chapter describes and illustrates the characteristics of contour lines and the features they represent. Detailed instructions are provided for constructing contour maps when given a minimum of elevation data.

The topics covered include:

- Contour line characteristics
- Types of contour lines
- Plotting contour lines from field notes
- Plotting contour lines with a CAD system
- Enlarging contour maps

CONTOUR LINE CHARACTERISTICS

The best example of a contour line is the shore of a lake or reservoir. The water level represents one contour line because the level of the lake is the same in all places. By late summer, many reservoirs are lowered considerably and previous water levels are seen as lines: contour lines. The space between these lines is termed the *contour interval*.

FIGURE 7–1. Contour lines formed by lapping water at different levels in a reservoir. *(Courtesy City of Portland, Oregon)*

If you observe the contour lines of a reservoir closely, you can see that they do not touch, and they run parallel to each other. One line can be followed all the way around the reservoir until it closes on itself. This basic characteristic is shown in Figure 7–1. Note also that the contour lines are generally parallel, and they never cross one another.

Slopes

The steepness of a slope can be determined by the spacing of the contour lines. A gentle slope is indicated by greater intervals between the contours (see Figure 7–2), whereas a steep slope is evident by closely spaced contours (see Figure 7–3).

Slopes are not always uniform. A concave slope flattens toward the bottom, as seen in Figure 7–4. The steepness of the upper part of the slope is represented by the close contour lines.

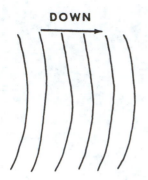

DOWN

FIGURE 7–2. Uniform gentle slope.

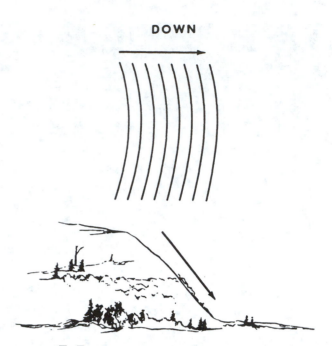

DOWN

FIGURE 7–3. Uniform steep slope.

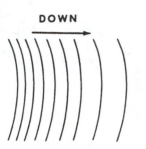

DOWN

CONCAVE

FIGURE 7–4. Concave slope.

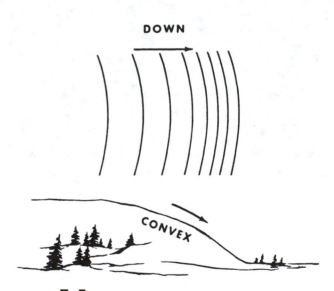

DOWN

CONVEX

FIGURE 7–5. Convex slope.

A *convex* slope is just the opposite and develops a steeper gradient as it progresses. The contour lines are closer together near the bottom of the slope, as seen in Figure 7–5.

An exception exists to the rule that contours never touch. In Figure 7–6, the contours are obviously touching. Can you determine the type of landform shown in Figure 7–6? It is a cliff. Sheer cliffs and vertical rock walls of canyons are easy to spot on topographical maps because of the sudden convergence of contour lines.

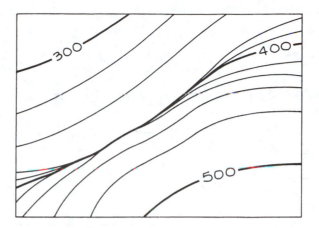

Streams and Ridges

Those of you who have taken hikes in the mountains know that as a level trail traverses a hill, you find yourself walking in a large "U" or horseshoe. As you approach a stream between two hills, the trail begins to form a "V" pointing upstream, as shown in Figure 7–7a. Near stream junctions, an "M" is often formed, as shown in Figure 7–7b. The peaks of the M point upstream.

The characteristics shown in Figure 7–7 can quickly reveal slope, stream flow, and slope directions to an experienced map

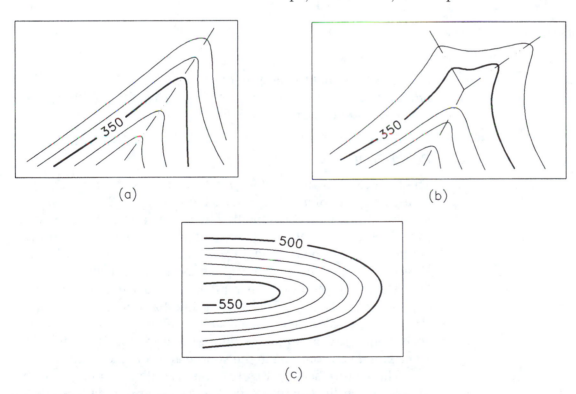

(a)   (b)

(c)

FIGURE **7–7.** (a) Contours form a "V" pointing upstream. (b) Contours form an "M" above stream junctions. The tops of the M point upstream. (c) Contours form a "U" around ridges. The bottom of the U points downridge.

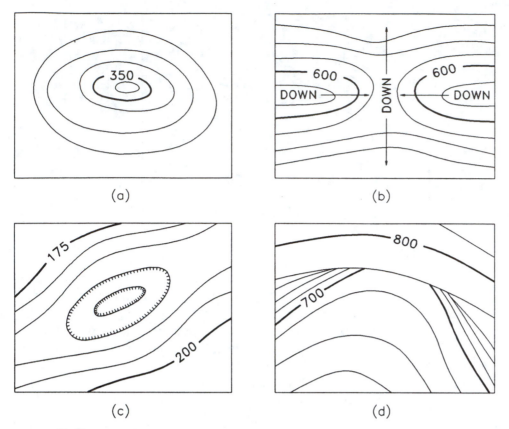

FIGURE 7–8. (a) Dome-shaped hill. (b) Saddle. (c) Depression. (d) Overhang.

reader. Notice that the bottom of the V points upstream. Contour lines form a U around a hill or ridge and the bottom of the U points downhill, as shown in Figure 7–7c.

Relief Features

The features shown in Figure 7–8 are easily spotted on contour maps. The peak of a hill or mountain in Figure 7–8a is a common feature and may be accompanied by an elevation. Here, the contour lines form circles of ever-decreasing diameter. Two high points or peaks side by side form a *saddle*, such as that shown in Figure 7–8b. A saddle is a low spot between two peaks.

The special contour line in Figure 7–8c is used to represent a depression in the land. The contour line has short lines pointing into the depression, and is used to identify human-made features such as quarries and pits, or natural features such as the limestone sinkholes common in the southeastern United States.

The feature seen in Figure 7–8d is similar to a cliff, but the lower contour lines actually seem to go under the higher ones. In reality they do, and portray an overhang, a feature found in rocky, highly sculpted, and mountainous terrain.

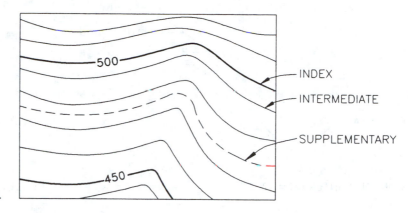

FIGURE 7–9. Types of contour lines.

TYPES OF CONTOUR LINES

Index Contours

Every fifth line on a topographic map is an index contour. This aids the map reader in finding references and even-numbered elevations. The index contour is normally a thick line and is broken at intervals and labeled with its elevation (see Figure 7–9).

Intermediate Contours

The remaining contours in Figure 7–9 are intermediate contours and represent the intervals of elevation between the index lines. There are four of these lines between index contours. These lines are not normally labeled, but can be if the scale and function of the map dictate.

Supplementary Contours

The supplementary contour is not as common as the others. It is used when the normal contour interval is too large to illustrate significant topographic features clearly on land with a gentle slope. They are usually given the value of half the contour interval (see Figure 7–9).

Determining Contour Line Values

When constructing contour maps, you will have to assign values to index contours. This should not be an arbitrary decision, but one based on a constant reference. That reference is sea level. Imagine that the first index contour is at mean sea level. *Mean* is the average elevation between high and low tides. Next there are four intermediate contours, then the fifth line is an index contour. Given the contour interval to be used for the map, it should be an easy process to determine the index contour values. Figure 7–10 provides examples of index contour values at a variety of contour intervals.

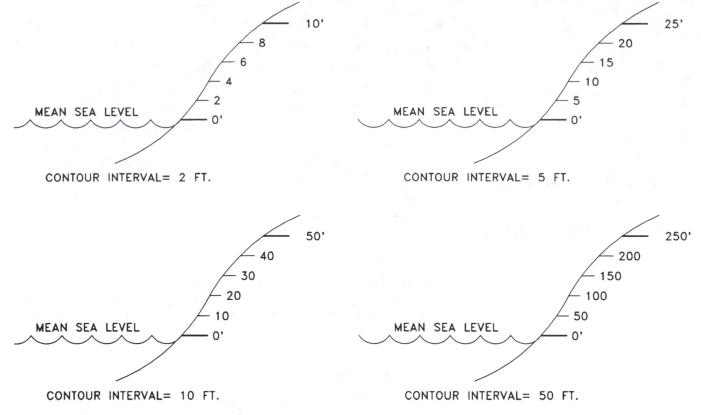

FIGURE **7–10.** Every fifth line is an index contour, and the contour interval determines the value of the index.

PLOTTING CONTOUR LINES FROM FIELD NOTES

As a rule, topographic maps tend to be somewhat inaccurate in representing the true shape of the land. General features and large landforms can be shown accurately, but small local relief is often eliminated. This is especially true of contour maps created from aerial photographs. Trees and vegetation may hide the true shape of the land.

A more accurate view of the land can be obtained by ground survey. This process of mapping establishes spot elevations from known points. Field notes are then plotted as contour lines on the new map. Let us take a closer look at this process.

Control Point Survey

Plots of land may often be surveyed in such a manner that contour maps can be created from the survey or field notes. The creation

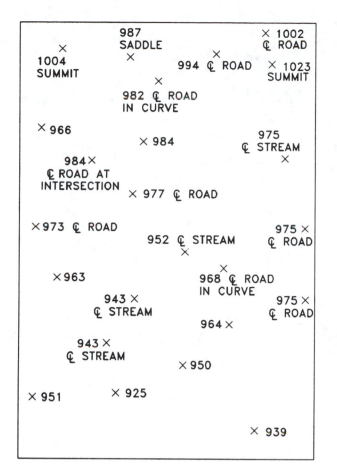

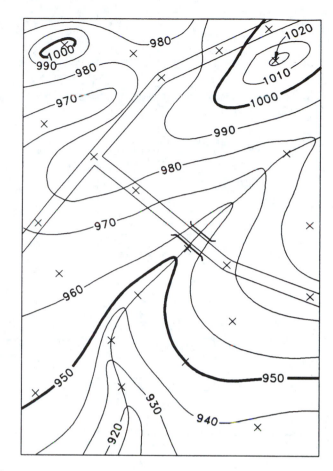

FIGURE **7–11.** Contour map plotted using control point survey.

of these maps depends on a certain amount of guesswork, but a good knowledge of landforms, slopes, roads, and stream characteristics may ensure a greater degree of success in the mapping operation.

The control point survey is a common method of establishing elevations for use in contour mapping. Figure 7–11 is an example of the two basic steps in the mapping process. The surveyor's field notes (elevations) are plotted on the map and labeled. Next a contour interval is chosen. This depends on the purpose of the map and the elevation differential within the plot. Contour lines are then drawn to connect equal elevation points. Notice in Figure 7–11 that prominent features on the landscape are surveyed. These are the control points. Their elevations may not be an even number. In this case, the mapper must use one of two methods to "interpolate" between two control points to find the even number. While we are on the subject, let us discuss interpolating a little further.

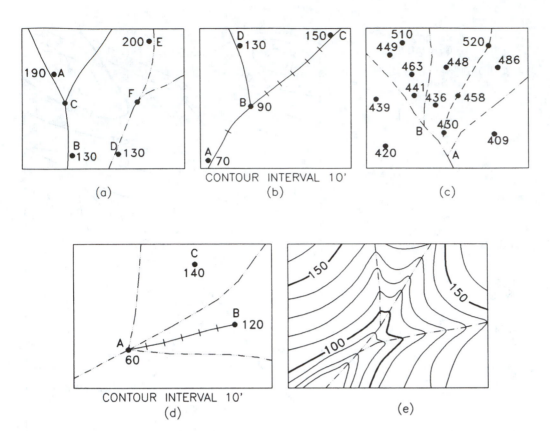

FIGURE 7–12. Interpolating contour lines.

Interpolating Contour Lines by Estimation

To interpolate means to insert missing values between numbers that are given. This is often the situation that the drafter faces when plotting field notes. Inserting missing values may seem like guesswork, but it can be accomplished with a certain amount of accuracy if five basic steps are followed. Only the surveyor or mapper with experience should use this method for establishing contour lines.

Step 1: Establish elevations at stream junctions. Given the elevations at points A and B in Figure 7–12a, we must determine the elevation of the stream junction at C. With dividers or scale, we find that C is two-thirds of the distance from A to B. The elevation difference between A and B is 60 ft. As ⅔ of 60 is 40, the elevation of point C is 170 ft. Using this method, determine the elevation of the stream junction at F.

Step 2: Locate points where contours cross streams. Given a contour interval of 10 ft in Figure 7–12b and spot elevations at points A, B, C, and D, we must locate stream crossings of all contour lines. The vertical elevation between points A and B is 20 ft. Divide the distance between A and B in half and the 80-ft contour can be located at that point. The vertical distance between points B and C is 60 ft. The line between these two points must be divided into six equal segments. The first new point above B is 100 ft in elevation. Determine the stream crossings of the contour lines between points B and D.

Step 3: Locate Ridges as Light Construction Lines. Areas of higher elevation usually separate streams except in swamps and marshes. To locate these ridges or *interfluves* (Figure 7–12c), sketch a light dashed line beginning at the stream junction and connecting the highest elevation points between the two streams. Sketch a similar line beginning at stream junction B.

Step 4: Determine the Points on the Ridge Lines Where Contours Cross. This step is performed in exactly the same manner as step 2. In Figure 7–12d, the space between point A and the spot elevation of 120 ft is 60 ft. The line is divided into six equal parts as in step 2. Locate the contour crossings between points A and C.

Step 5: Connect the Points of Equal Elevation with Contour Lines. As you do this, keep in mind the characteristics of contour lines discussed previously (see Figure 7–12e).

Be aware that interpolating distances is necessary in situations other than those discussed in the five steps. Keep in mind that by interpolating contour lines, we are using an assumption termed *uniform slopes*. This is based on equal spacing of contours between known points. Some do's and don'ts are illustrated in Figure 7–13.

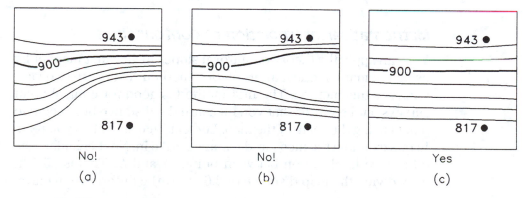

| No! | No! | Yes |
| (a) | (b) | (c) |

FIGURE 7–13. When interpolating contour lines using the "uniform slopes" concept, always space contours evenly as in example "c."

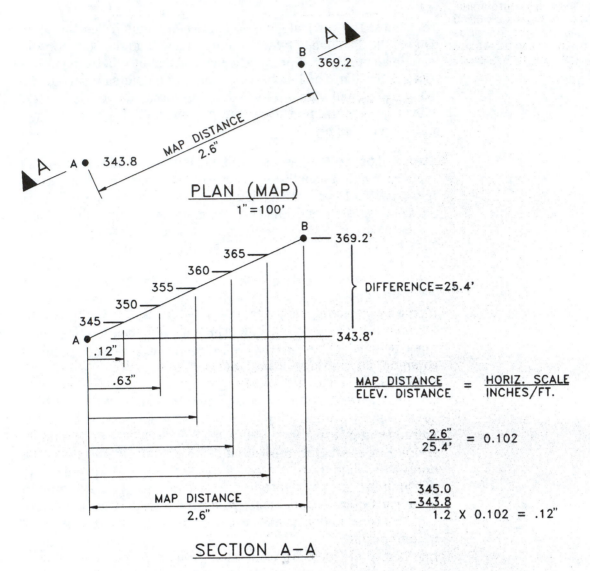

PLAN (MAP)
1"=100'

MAP DISTANCE
2.6"

MAP DISTANCE
2.6"

DIFFERENCE=25.4'

$$\frac{\text{MAP DISTANCE}}{\text{ELEV. DISTANCE}} = \frac{\text{HORIZ. SCALE}}{\text{INCHES/FT.}}$$

$$\frac{2.6"}{25.4'} = 0.102$$

345.0
−343.8
1.2 × 0.102 = .12"

SECTION A—A

FIGURE 7–14. Mathematical interpolation of contour lines allows you to calculate the distance between two points.

Mathematical Interpolation of Contour Lines

You can apply the theory of uniform slopes in an exact manner by using the mathematical interpolation method. There are two variations of this method. The first method is good for control point surveys, and requires that you measure the distance between two points using the scale of the map. Look at Figure 7–14 as you read. In this example the measured distance is 2.6 in. Next calculate the difference in elevation between points *A* and *B*. This is 25.4 ft. Now divide the map distance of 2.6 in. by the difference in elevation, 25.4 ft, to get the horizontal scale distance in inches per foot of vertical rise.

$$\frac{\text{Map distance}}{\text{Elevation difference}} = \text{in./ft of vertical rise}$$

$$\frac{2.6 \text{ in.}}{25.4 \text{ ft.}} = 0.102 \text{ in./ft of vertical rise}$$

If the contour interval is 5 ft, the first contour above point A is 345 ft. The difference between this contour and point A is 1.2 ft. Multiply (1.2 × 0.102), and the result is 0.12 in. This is the distance that you would measure from point A on the map to find the 345-ft contour line.

Continue in this manner to locate the 350 foot contour on the map. The difference between 350 and the 343.8 value of point A is 6.2. Multiply this by 0.102 and the result is 0.63 ft. This is the distance you would measure from point A on the map to find the 350-ft contour line. Find the remaining contours using this same method and record your answers in the spaces provided in Figure 7–14.

The second method uses the map distance between two points to find a percent of slope. The slope can then be converted to an actual map distance from the given elevation to contour lines. Look at Figure 7–15 as you follow this explanation. The elevation difference between points A and B is 25.4 ft. Divide 25.4 ft by the map distance of 260 ft.

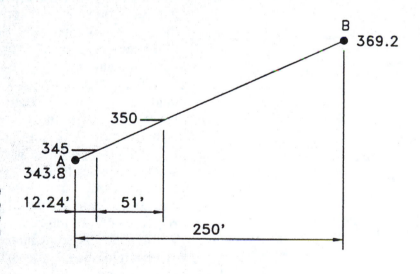

FIGURE **7–15.** The map distance between two points can be used to determine the percent of slope. The slope is converted to a distance between a given elevation and a contour line.

$$25.4 \text{ ft} \div 260 \text{ ft} = 0.098$$

This gives a percent of slope of 0.098. Next, subtract the elevation of point A from the nearest contour, 345. To solve for the distance from point A to the 345-ft contour (x), divide 1.2 by 0.098.

$$0.098x = 1.2 \text{ ft}$$

$$x = 12.24 \text{ ft}$$

The result of 12.24 ft is the distance to measure on the map from point A to find the 345-ft contour. Use this same method to solve for the location of each additional contour from point A.

Another variation of mathematical interpolation is best used for grid surveys, and is discussed in the next section of this chapter.

Grid or Checkerboard Surveys

Using a grid, the surveyor divides the plot of land into a checkerboard. Stakes are driven into the ground at each grid intersection. Elevations are measured at each stake and recorded in the field book. Additional stakes may be placed and recorded if significant elevation changes occur between grid intersections. The drafter plots this grid as shown in Figure 7–16a. The size of the squares is determined by the surveyor based on the land area, topography, and elevation differential. The vertical lines of the grid in Figure 7–16 are labeled with letters and are 20 ft apart. The horizontal lines are labeled 0 + 00, 0 + 20, 0 + 40, 0 + 60, and so on, and are called *stations*. The stations are also 20 ft apart. The first number in the station label is hundreds of feet, and the second number to the right of the plus sign (+) is tens of feet. For example, the station number 5 + 45 is 5 hundreds and 45 tens, or 545 ft from the point of beginning.

The elevation of each grid intersection is recorded in the field notes shown in Table 7–1. Using the surveyor's field notes, we see that grid point A–0 + 40 has an elevation of 100.0 ft. What is the elevation of point C–0 + 60?

To begin plotting the elevations, draw a grid at the required scale, or use grid paper. Next, using the field notes, label all the grid intersections with elevations (see Figure 7–16b). Connect the elevations with freehand lines. The drafter is aware of the required contour interval at this point and must decide which elevation points to connect and when to interpolate between uneven elevation points (Figure 7–16c). Figure 7–16d shows the finished contour map.

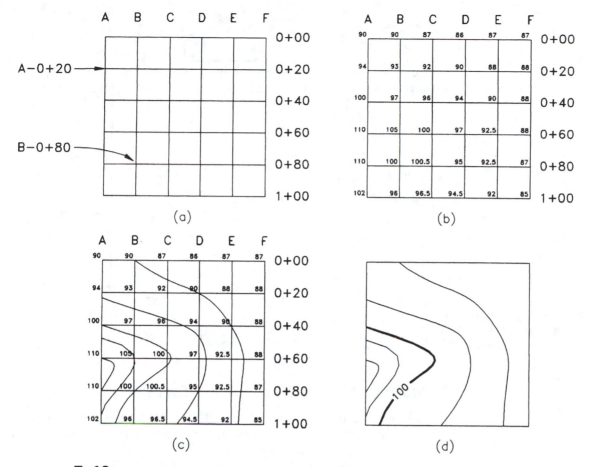

FIGURE 7–16. (a) For a grid survey, land is divided into a checkerboard and labeled. (b) All grid intersections are labeled. (c) Connect the elevations with freehand lines. (d) Completed contour map of grid survey.

TABLE 7–1. Grid survey field notes for the map in Figure 7–16

Station	Elev.	Station	Elev.	Station	Elev.
A–0 + 00	90.0	C–0 + 00	87.0	E–0 + 00	87.0
A–0 + 20	94.0	C–0 + 20	92.0	E–0 + 20	88.0
A–0 + 40	100.0	C–0 + 40	96.0	E–0 + 40	90.0
A–0 + 60	110.0	C–0 + 60	100.0	E–0 + 60	92.5
A–0 + 80	110.0	C–0 + 80	100.5	E–0 + 80	92.5
A–1 + 00	102.0	C–1 + 00	96.5	E–1 + 00	92.0
B–0 + 00	90.0	D–0 + 00	86.5	F–0 + 00	87.0
B–0 + 20	93.0	D–0 + 20	90.0	F–0 + 20	88.0
B–0 + 40	97.0	D–0 + 40	94.0	F–0 + 40	88.0
B–0 + 60	105.0	D–0 + 60	97.0	F–0 + 60	88.0
B–0 + 80	100.0	D–0 + 80	95.0	F–0 + 80	87.5
B–1 + 00	96.0	D–1 + 00	94.5	F–1 + 00	85.0

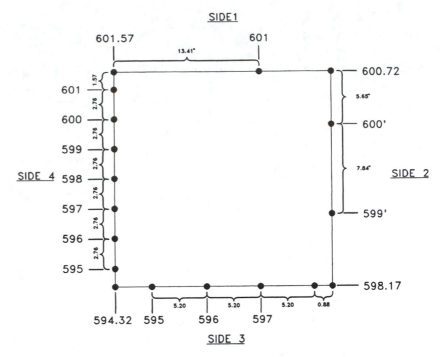

Interpolating Contours for Grid Surveys

A mathematical interpolation method can be used to plot contours in a grid survey using the theory of uniform slopes. Look at Figure 7–17a as you follow this example.

The formula requires three values: (1) the amount over the whole foot contour of the highest point you are working from; (2) the difference in elevation between the point you are working from and the next lowest grid elevation; (3) the distance between the point you are working on and the next lowest grid intersection. The formula is written as follows:

$$\frac{\text{Amount over whole foot contour}}{\text{Difference in elevation}} \times \text{Distance}$$

Use the formula to calculate where the 601-ft contour will fall along side 1 in Figure 7–17a. The following examples begin calculating at the highest elevation of the two grid intersections in question. (The number 20 in the formula below represents the distance between grid intersections.)

$$\text{Side 1:} \quad \frac{0.57}{0.85} \times 20 = 13.41$$

Measure 13.41 ft from the 601.57-ft point to find the 601-ft contour. Calculate side 2 using the following formulae:

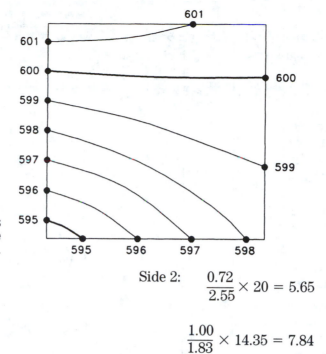

FIGURE **7–17b.** After elevation values are located on the grid, points of the same value are connected to form contour lines.

Side 2: $\dfrac{0.72}{2.55} \times 20 = 5.65$

$$\dfrac{1.00}{1.83} \times 14.35 = 7.84$$

Notice that only two calculations are required per side if more than two contours cross a side. The second formula calculates the horizontal distance for 1 ft of elevation. Therefore no additional calculations are needed. The same process is used for the third and fourth sides.

Side 3: $\dfrac{0.17}{3.85} \times 20 = 0.88$

$$\dfrac{1.00}{3.68} \times 19.12 = 5.20$$

Side 4: $\dfrac{0.57}{7.25} \times 20 = 1.57$

$$\dfrac{1.00}{6.68} \times 18.43 = 2.76$$

After you have located the contour line elevations on the edges of the grid, connect points of the same value with lines, as shown in Figure 7–17b.

Radial Survey

The *radial survey* is a technique that can be used for locating property corners, structures, natural features, and points to be

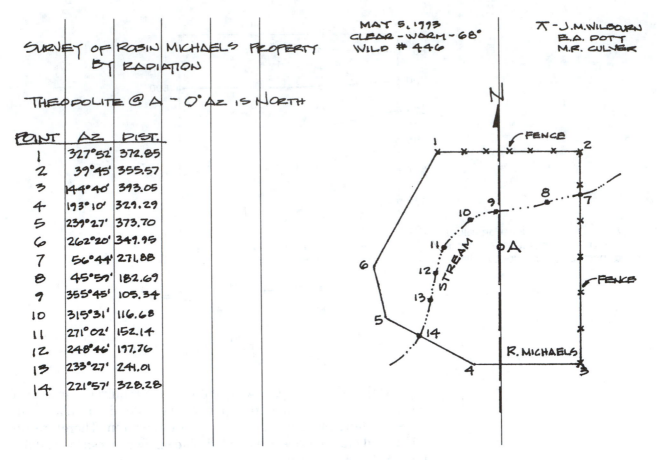

The handwritten field notes contain:

SURVEY OF ROBIN MICHAELS PROPERTY BY RADIATION

THEODOLITE @ A - 0° AZ IS NORTH

POINT	AZ	DIST.
1	327°52'	372.85
2	39°45'	355.57
3	144°40'	393.05
4	193°10'	329.29
5	239°27'	373.70
6	262°20'	349.95
7	56°44'	271.88
8	45°57'	182.69
9	355°45'	105.34
10	315°31'	116.68
11	271°02'	152.14
12	248°46'	197.76
13	233°27'	244.01
14	221°57'	328.28

MAY 5, 1993
CLEAR - WARM - 68°
WILD # 446

T - J.M. WILBOURN
B.A. DOTT
M.R. CULVER

FIGURE 7–18. Field notes of a radial survey contain azimuths and distances property corners and additional control points.

used in contour mapping. Control points are established by a method called *radiation,* in which measurements are taken from a survey instrument located at a central point, called a *transit station.* From this point a series of angular and distance measurements are made to specific points. Readings such as this, taken from a transit station to specific control points, are called *side shots.* This technique is also used when an instrument setup (station) cannot be located on a point such as a property corner or section marker.

From the transit station, the azimuths and distances of each point are recorded in the field notes. See Figure 7–18. This information can later be used to construct a property plat or a contour map. If the property is too large to sufficiently map using a single transit station, additional transit stations can be established. When two transit stations are used, the line between the two is called a *base line.* When more than two stations are used, the lines between them are called *transit lines.* Several transit lines can form an open or closed traverse.

FIGURE **7–19.** A property plat can be constructed from the radial survey field notes.

PROPERTY PLAT

SCALE: 1" = 200'

Any of the interpolation techniques previously discussed can be used to construct a contour map. The field notes in Figure 7–18 were used to create the map shown in Figure 7–19.

Contour Labeling

Most topographic maps show written elevations only on index contour lines. These labels are normally enough to give the map reader sufficient reference values to work with. Figure 7–20 shows the method in which labeling is done. The elevation numbers are placed so that they are not upside down. Contour line labels are located at regular intervals along the contour line.

Property plats, highway maps, and special maps of many kinds may require every contour to be labeled. In this case the

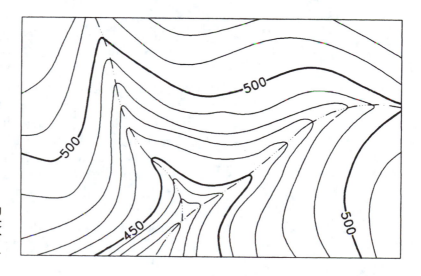

FIGURE **7–20.** Contour line labeling should be on index contours, placed at regular intervals, and should not appear upside down.

drafter must use good layout and spacing techniques to achieve a balanced and uncluttered appearance.

PLOTTING CONTOUR LINES WITH A CAD SYSTEM

If a CAD drafter begins with a grid survey, such as the one shown in Figure 7–16, contour lines can be interpolated and drawn in a manner similar to the one previously mentioned. The grid, of course, would be drawn in actual feet instead of to a particular scale. Also, CAD software packages will use the stationing and elevations given, then interpolate and draw the contour lines automatically.

Index contours can also be labeled automatically with CAD. The drafter usually has options about the height of lettering and the spacing of lettering along the contours. Again, the drafter must use good judgment with these options and not rely solely on the capabilities of CAD to spontaneously produce a balanced drawing.

ENLARGING CONTOUR MAPS

The civil drafter may be required to enlarge contour maps to show greater cultural detail or to define the topography with additional contour lines. Other than the photographic process or using a scaling command in a CAD system, the grid system is the best drafting method to use when manually enlarging (or reducing) a map.

Grid Layout

First, a grid must be drawn on the existing map. The size of the grid squares depends on the complexity of the map and the amount of detail you wish to show. You may be instructed to enlarge the map to twice its present size, or a scale for the new map will be specified. If you draw the grid on the existing map with ¼-in. squares and wish to double the size, the grid for your new map is drawn with ½-in. squares. Suppose that the scale of the original map is 1 in. = 1000 ft and the new map is to be 1 in. = 250 ft. If you draw the grid on the original using ¼-in. squares, the new grid is then drawn using 1-in. squares—four times the size.

The map in Figure 7–21 must be enlarged to twice its size and has been overlayed with a ¼-in. grid. Notice, too, that the vertical and horizontal grids have been numbered to avoid confusion when transferring points. Keep in mind that when a linear measurement is doubled, the area of that map is increased four times the original size. The appearance of Figure 7–22 illustrates this point.

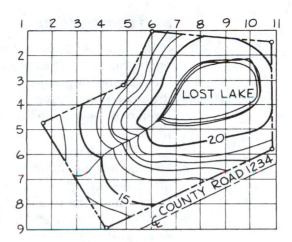

FIGURE 7–21. Grid drawn over original map.

The grid for the new map is constructed using the same number of squares as the original and is labeled the same. The only difference is its size (see Figure 7–22). The squares now measure ½ in.

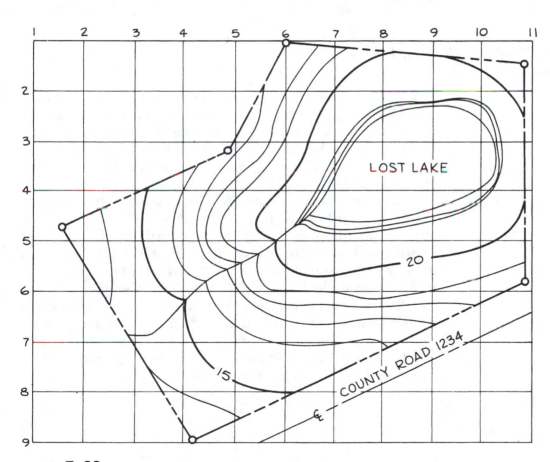

FIGURE 7–22. Enlarged map uses same number of grid lines, but dimensions of squares are doubled.

Map Construction

Map features can now be transferred from the original to the enlargement by eye, engineer's scale, or proportional dividers. Measurements provide the greatest degree of accuracy, but the drafter soon realizes that "eyeballing" features on or near grid intersections may be accurate enough for the purposes of the map.

The two grids that you draw should be exactly the same, as well as the labeling you use to identify the vertical and horizontal grid lines. Compare Figures 7–21 and 7–22. The process of transferring from the original to the enlargement is relatively simple. Choose one square and measure or estimate where features touch or cross the grid lines of that square. Transfer that measurement to the same square on the enlarged grid, remembering, of course, to increase its size proportionately.

The grid enlargement method can produce accurate maps provided that the drafter establishes a proper coordinate system, uses good measuring techniques, and avoids unnecessary "artistic license."

ENLARGING AND DIGITIZING CONTOUR MAPS WITH A CAD SYSTEM

Since CAD drawings are produced in real world measurements, enlarging a map is "simply" a matter of giving a different scale to the drawing when it is plotted. Care must be taken, however, to ensure that text height, spacing, and other characteristics are appropriate to the plotted drawing. CAD text height can often be changed with one command.

Manually drafted contour drawings can also be transferred to CAD by a method called *digitizing*. A scaled contour drawing is placed on a digitizer, the digitizer is set to the same scale as the drawing, and points are picked along the contour lines by a puck or mouse. These points are then part of the CAD drawing in real world measurements, and can be plotted at any scale.

Digitizing has many uses. For example, a CAD operator can digitize drainage areas of land in a city from existing city maps. This is for the purpose of determining how much storm water runoff will occur there. Engineers can then design storm sewers that will adequately handle storm water.

TEST

7-1 What do contour lines represent?

7-2 List four characteristics of contour lines.

7-3 Sketch the following features using contour lines:

Mountain peak   Saddle   Depression

7-4 What is the function of index contours?

7-5 Briefly define control point survey.

7-6 What is interpolation?

7-7 What type of survey divides the land into a checkerboard?

7-8 Explain the grid system of map enlarging.

7-9 What letter of the alphabet is formed when a contour line crosses a stream? _____ In which direction does the letter point?_____

7-10 What letter is formed when a contour line wraps around a hill? _____ In which direction does the letter point?_____

7-11 What is a radial survey?

7–12 What measurements are taken in a radial survey?

7–13 Explain how a manually drawn contour map could be transferred to a CAD system.

PROBLEMS

P7–1 Using proper methods of contour line interpolation, establish contour lines for the problems given in Figure P7–1. Use indicated contour intervals. Label all index contours and keep in mind the characteristics of contours discussed in this chapter.

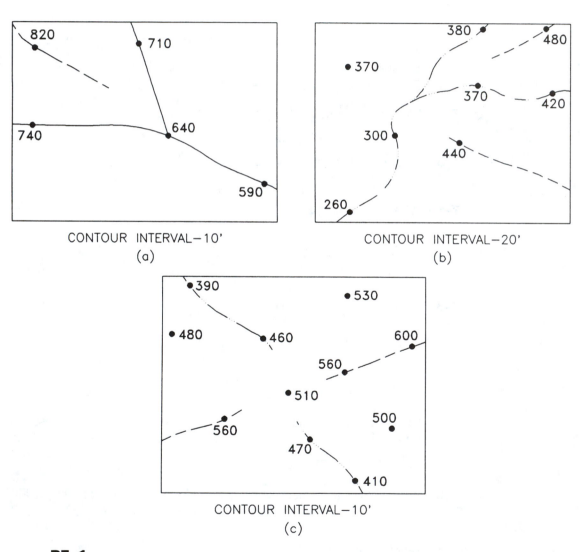

CONTOUR INTERVAL—10'
(a)

CONTOUR INTERVAL—20'
(b)

CONTOUR INTERVAL—10'
(c)

FIGURE **P7–1.**

P7–2 a. Given the surveyor's field notes from a control point survey, establish contour lines at the contour interval indicated in Figure P7-2. Index contours should be a heavy line weight and should be labeled. Use interpolation where required.

b. Assume the drawing shown in Figure P7–2 is at a scale of 1 in. = 200 ft. With your CAD system, digitize the drawing, also transferring all the elevation points. Interpolate and construct contours within CAD at an interval of 5 ft. Index contours should be a heavy line weight and should be labeled. Place labels of road, driveway, and other features where appropriate, and give them a finished plotted height of ⅛ in. Elevation labels should have a finished plotted height of ⅛ in. Plot at 1 in. = 50 ft on a D-size sheet of paper, and submit to the instructor for approval.

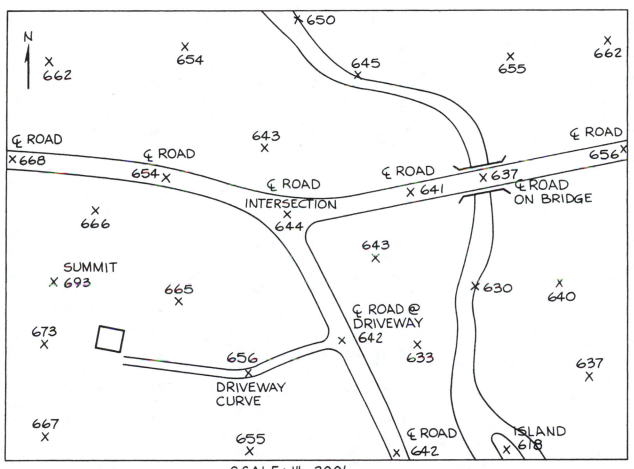

FIGURE **P7–2.**

P7–3   a. A grid survey produced the field notes given in Table P7–1. Using the grid of Figure P7–3 (shown on page 191), first locate the elevations of all grid intersections. Then plot contour lines at an interval of 10 ft. Index lines should be labeled and contrast in line weight with intermediate contours. Use a scale of 1″ = 100′.

b. With your CAD system, construct a grid similar to the one in Figure P7–3, labeling all station points. Use the information given in Table P7–1 to then locate the elevations of all grid intersections. Draw the contour lines at an interval of 10 ft. Index lines should be labeled and should contrast in line weight with intermediate contours. Plot at a scale of 1 in. = 200 ft, and submit to the instructor for evaluation.

TABLE P7–1.   Grid survey field notes

Station	Elev.	Station	Elev.	Station	Elev.
A–0 + 00	592	D–0 + 00	577	G–0 + 00	602
A–1 + 50	595	D–1 + 50	536	G–1 + 50	592
A–3 + 00	599	D–3 + 00	531	G–3 + 00	561
A–4 + 50	583	D–4 + 50	519	G–4 + 50	529
A–6 + 00	560	D–6 + 00	468	G–6 + 00	460
A–7 + 50	558	D–7 + 50	475	G–7 + 50	380
A–9 + 00	577	D–9 + 00	492	G–9 + 00	395
A–10 + 50	589	D–10 + 50	496	G–10 + 50	422
A–12 + 00	594	D–12 + 00	498	G–12 + 00	437
B–0 + 00	587	E–0 + 00	579	H–0 + 00	584
B–1 + 50	600	E–1 + 50	562	H–1 + 50	568
B–3 + 00	648	E–3 + 00	535	H–3 + 00	536
B–4 + 50	594	E–4 + 50	507	H–4 + 50	507
B–6 + 00	537	E–6 + 00	450	H–6 + 00	441
B–7 + 50	543	E–7 + 50	437	H–7 + 50	381
B–9 + 00	561	E–9 + 00	463	H–9 + 00	372
B–10 + 50	563	E–10 + 50	465	H–10 + 50	406
B–12 + 00	565	E–12 + 00	464	H–12 + 00	427
C–0 + 00	571	F–0 + 00	602	I–0 + 00	555
C–1 + 50	576	F–1 + 50	586	I–1 + 50	537
C–3 + 00	563	F–3 + 00	560	I–3 + 00	513
C–4 + 50	578	F–4 + 50	532	I–4 + 50	483
C–6 + 00	500	F–6 + 00	461	I–6 + 00	442
C–7 + 50	518	F–7 + 50	394	I–7 + 50	382
C–9 + 00	536	F–9 + 00	428	I–9 + 00	359
C–10 + 50	535	F–10 + 50	444	I–10 + 50	391
C–12 + 00	534	F–12 + 00	451	I–12 + 00	417

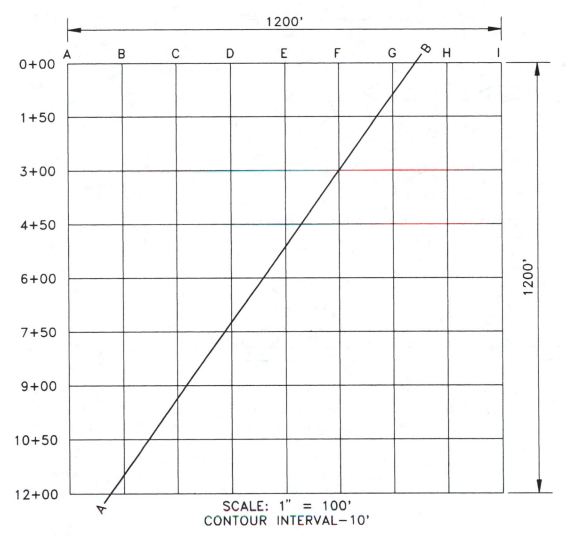

1200'

1200'

SCALE: 1" = 100'
CONTOUR INTERVAL—10'

FIGURE P7–3.

P7–4 Construct a grid enlargement of Figure P7–2 or P7–3. Increase the scale four times. Label all contours. Include a north arrow and legend. Use a C-size sheet of vellum or Mylar. Either pencil or ink can be used (check with instructor). Make a diazo print and submit it to the instructor for evaluation, unless otherwise indicated.

P7–5 Identify the features shown below.

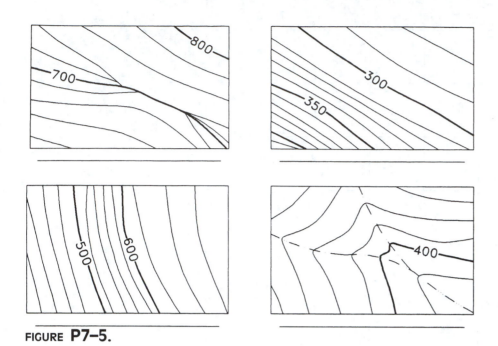

FIGURE **P7–5.**

P7–6 A radial survey produced the site notes shown in Figure P7–6. Using the elevation points given, create a contour map of the site with a contour interval of 2 ft. In the completed contour plan, show all set nail locations and elevations, as well as the road bed and concrete meter vault. The side cut on the north side of the gravel road should also be represented. Label only index contour lines. Include a north arrow, verbal scale, and graphic scale.

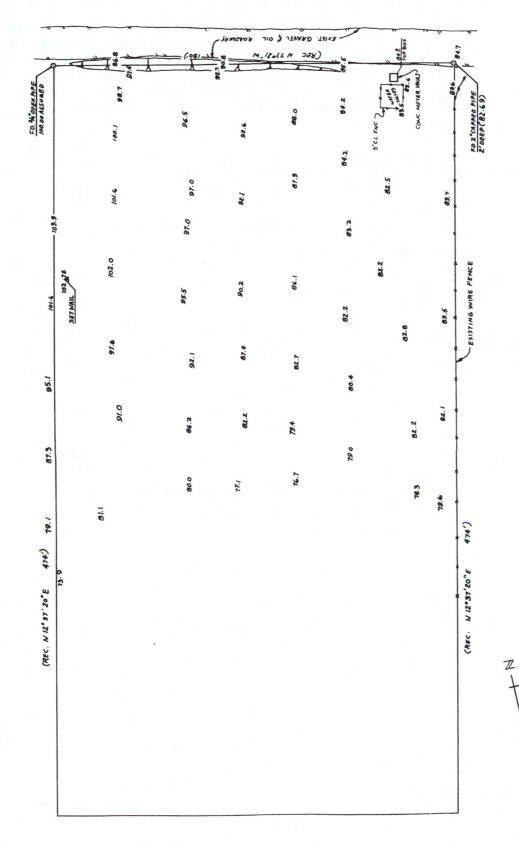

FIGURE **P7-6.**

P7–7 A radial survey produced the site notes shown in Figure P7–7. Using the elevation points given, create a contour map of the site with a contour interval of 5 ft. Draw the map using a scale of 1 in. = 20 ft. Locate the elevation points by measuring in Figure P7–7. The scale of Figure P7–7 is 1 in. = 60 ft. Label all features indicated, and label all contour lines. Include a north arrow, and a verbal scale.

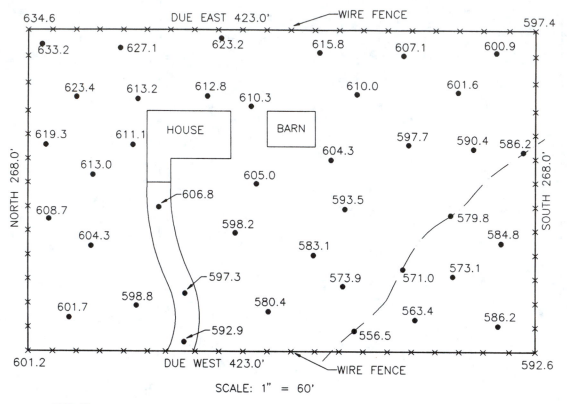

FIGURE **P7–7.**

P7–8 A radial survey produced the field notes in Figure P7–8. Use these notes to construct a site map with a contour interval of 1 ft. Label all of the control points and each index line. Show all features indicated in the field notes. Include a north arrow, representative fraction scale, and graphic scale.

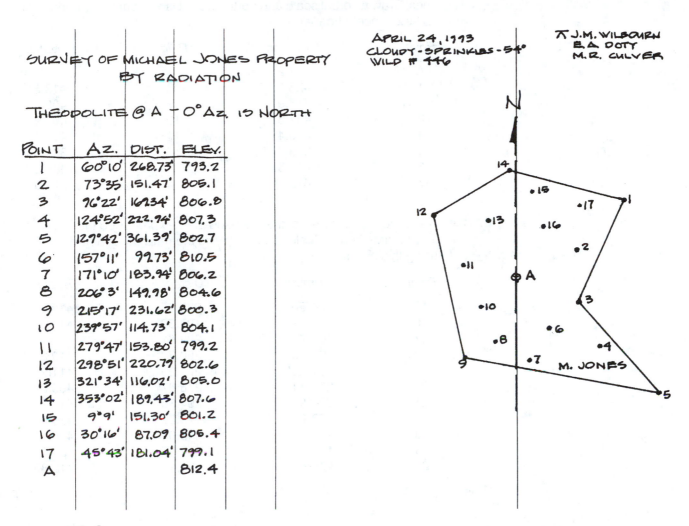

SURVEY OF MICHAEL JONES PROPERTY BY RADIATION

THEODOLITE @ A — 0° Az. IS NORTH

POINT	AZ.	DIST.	ELEV.
1	60°10'	268.73'	793.2
2	73°35'	151.47'	805.1
3	76°22'	162.34'	806.8
4	124°52'	222.74'	807.3
5	127°42'	361.39'	802.7
6	157°11'	92.73'	810.5
7	171°10'	183.94'	806.2
8	206°3'	149.78'	804.6
9	215°17'	231.62'	800.3
10	239°57'	114.73'	804.1
11	279°47'	153.80'	799.2
12	298°51'	220.77'	802.6
13	321°34'	116.02'	805.0
14	353°02'	189.43'	807.6
15	9°9'	151.30'	801.2
16	30°16'	87.09'	805.4
17	45°43'	181.04'	799.1
A			812.4

APRIL 24, 1773
CLOUDY - SPRINKLES - 54°
WILD # 446

⊼ J.M. WILBOURN
E.A. DOTY
M.R. CULVER

FIGURE P7–8.

P7–9   Refer to Figure 7–18 for this problem. Plot the piece of property using the field notes given in Figure 7–18. Draw the stream as intermittent. Construct a contour map of the property using the elevations of each of the following points. These points correspond to the points in the field notes. Use a contour interval of 2 ft, and label the index contour lines. Draw a verbal scale and north arrow. Use a scale appropriate for a B-size sheet of paper.

Point	Elevation	Point	Elevation
A	421.2		
1	423.5	8	413.6
2	418.9	9	409.1
3	437.3	10	407.9
4	426.4	11	401.3
5	389.7	12	394.6
6	412.5	13	386.5
7	416.2	14	372.8

P7–10   A second radial survey of the site used in P7–9 produced the following list of points and their elevations. Revise the map constructed in P7–9 using this new survey.

Point	Azimuth	Distance	Elevation
15	20°50′	228.75′	418.3′
16	344°59′	197.30′	419.6′
17	302°48′	282.61′	416.7′
18	291°10′	200.73′	410.2′
19	271°20′	271.39′	411.9′
20	250°11′	276.97′	406.0′
21	204°2′	204.65′	414.7′
22	172°33′	116.33′	423.5′
23	149°15′	246.61′	428.1′
24	113°35′	248.07′	427.0′
25	78°40′	123.11′	412.8′

CHAPTER 8

Profiles

A *profile* is an outline. An *artistic profile* is the outline of a face from the side, and a *map profile* is the outline of a cross section of the earth. Profiles are drawn using the information given on contour maps. Their uses include road grade layout, cut-and-fill calculations, pipeline layouts, site excavations, and dam and reservoir layout. This chapter examines basic profile construction from contour maps and the *plan and profile* commonly used by civil engineering firms for underground utility location and layout and for highway designs.

This chapter discusses the purposes and types of map profiles. Instructions and illustrations are provided to show how map profiles can be used in a variety of situations.

The topics covered include:

- Contour map profiles
- Profile leveling
- Plan and profile
- Plan and profile with CAD

CONTOUR MAP PROFILES

Constructing a contour map profile is a simple process, and is used if a profile leveling survey (discussed in the next section) has not yet been conducted. The only resource needed for the contour map profile is the map itself.

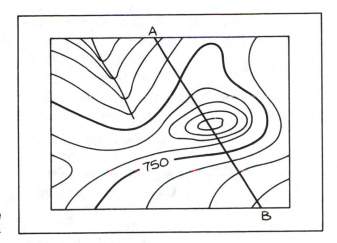

FIGURE **8–1.** Profile to be cut along line *AB*.

Map Layout

A straight line should be drawn on the map where the profile is to be made, as shown in Figure 8–1. The line between points *A* and *B* may be a proposed road or sewer line, and it may be at an angle other than horizontal on your drawing board. For ease of projec-

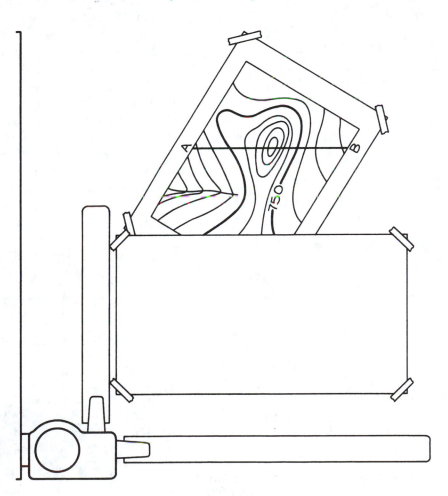

FIGURE **8–2.** Correct relationship of map, drawing paper, and drafting machine for profile construction.

tion, turn the map so that the profile line is horizontal and aligned with the horizontal scale of your drafting machine. A clean sheet of paper can then be placed directly below the profile line and used to construct the cross section (see Figure 8–2).

Profile Construction

The horizontal scale of the profile is always the same as the map because the profile is projected from the map. The vertical scale may be exaggerated to give a clear picture of the shape of the land. The amount of exaggeration depends on the relief of the map, the scale of the map, and the purpose of the profile.

The length of the profile is established by projecting end points A and B to your paper. The height of your profile depends first on the amount of relief between points A and B. Find the lowest and highest contours that line AB crosses and subtract to determine the total amount of elevation to be shown in the profile. This enables you to establish a vertical scale to fit the paper and best show the relief.

The vertical scale of a profile is exaggerated to show the elevation differences. In most cases, if you used the horizontal scale for the vertical, there would appear to be little change in elevation. Choose a vertical scale that fits the space allotted for the drawing, or use the 10:1 ratio that is often used in civil engineering. For example, if the horizontal scale is 1 in. = 100 ft, the vertical scale would be 1 in. = 10 ft.

When constructing the vertical profile scale, provide an additional contour interval above and below the extreme points of the profile. Also notice in Figure 8–3 that the elevations are labeled along one side and the scales are written by the profile. The vertical scale is sometimes written vertically near the elevation values.

Project horizontal lines from the vertical scale values across the drawing, then project all points from the map where the pro-

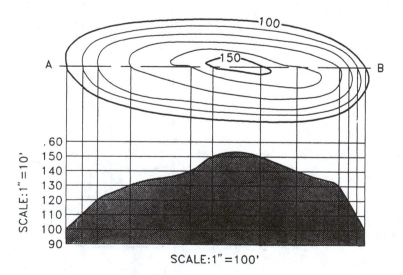

FIGURE 8–3. Layout and construction of map profile.

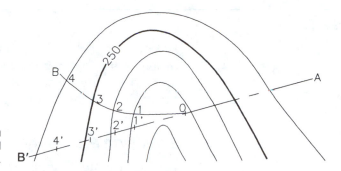

FIGURE 8–4. Profile construction of a curved line. Measure distances on *AB* and establish them on straight line *AB'*.

file line crosses contour lines. Notice in Figure 8–3 that a point on the profile is established where vertical and horizontal lines of the same elevation intersect. Once all these points are established, connect them with a freehand line. A sectioning symbol or shading is used to indicate the ground that is cross-sectioned.

Profiles of Curved Lines

Layout of a profile from a straight line is simple, but plotting a profile from a curved line involves an additional step. Before the profile can be drawn, curved line *AB* must be established along a related straight line, *AB'*. This can be done with dividers, compass, or engineer's scale (see Figure 8–4). Label the new points on line *AB'* to avoid any confusion. Note that the straight line distance 0-1 is transferred to 0-1', 1-2 to 1'-2' and so on.

From this point, creating the profile is the same as it is for a straight line. Be certain that you project the actual contour crossings from point *A* to 0 and the newly established points, 1' to 4', from 0 to *B'*. See Figure 8–5 for the proper method of projection.

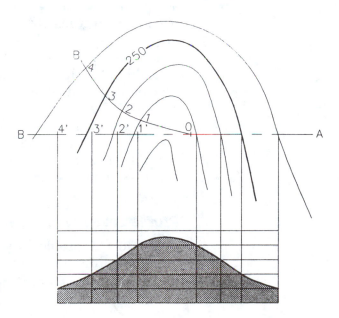

FIGURE 8–5. Construct profile of curved line *AB* from new straight line *AB'*.

Before a project such as a highway, utility line, railroad, canal, or other linear feature can be built, the elevation characteristics along the route must be surveyed. In *profile leveling*, surveyors measure a number of elevations along the centerline of the route.

Surveying Procedure

Profile leveling is similar to the standard differential leveling, but additional intermediate shots are taken along the route. The instrument is set up in a position from which several shots along the centerline can be taken. The rod is placed on a known elevation, such as a B.M., and a backsight (B.S.) is taken to determine the height of the instrument (H.I.). The rod can then be placed at intervals along the route's centerline and elevations measured at each point. These rod shots are called *intermediate foresight readings (IFS)*, or *ground rod readings*.

The location of IFS readings is determined by the topography. Normally, stakes are set at 50- or 100-ft intervals. Stations set at 100-ft intervals are called *full stations*. They are labeled 0 + 00, 1 + 00, 2 + 00, and so on. Intermediate readings taken between full stations are referred to as *plus stations*. For example, a point that is 463.25 ft from station 0 + 00 is labeled 4 + 63.25. Plus stations are located at sudden elevation changes such as bank and stream edges, road edges and centerlines, and the tops and bottoms of ditches.

When IFS readings can no longer be taken from the instrument setup, a new instrument station is needed. The rod is placed on a turning point (T.P.), and a foresight (F.S.) reading is taken. The instrument is then moved ahead to a location from which additional IFS readings can be made. Figure 8–6 shows an example of what a typical profile leveling setup would look like on a map.

To check the profile leveling survey, it is necessary to close on another BM. If you remember from the leveling discussion in Chapter 2, the difference between the totals of the B.S. readings and the F.S. readings is the elevation difference between the beginning and ending points. For example, if the totals of the B.S. readings are + 14.65, and the totals of the F.S. readings are −6.79, the difference is + 7.86 (+14.65 −6.79 = +7.86). This value should be the H.I. reading when the rod is placed on the final BM.

Profile Leveling Field Notes

Field notes for profile leveling are similar to those of differential leveling, with the addition of station values and IFS readings. Figure 8–7 shows an example of profile leveling field notes.

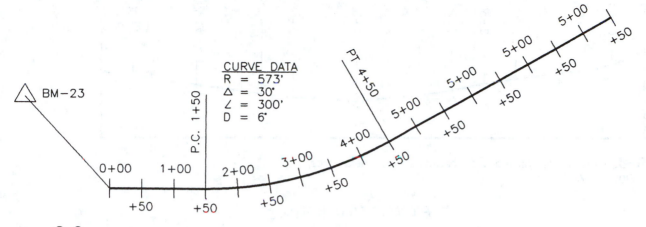

FIGURE 8-6. Profile leveling measurements are taken at the station points indicated on the map.

PROFILE LEVELING
PROPOSED HOLCOMB BIKE PATH

JULY 22, 1973
SUNNY-WARM-81°
TOPCON AUTO LEVEL #8148

7-J.M. WILBOURN
ROD- B.A. DOTT
NOTES- L.J. ROBINSON

STA	BS	HI	FS	IFS	ELEV	
BM-23	4.12	319.12			315.00	BRASS MONUMENT 198' NW OF STA. 0+00
0+00				4.99	314.13	
+50				6.77	312.35	
1+00				7.31	311.81	
+50				7.95	311.17	
2+00				8.66	310.46	
TP-1	3.01	313.18	8.95		310.17	STONE
2+50				3.37	309.81	
3+00				4.58	308.60	
+50				5.35	307.83	
4+00				5.71	307.47	
+50				6.48	306.70	
5+00				5.92	307.26	
+50				5.70	307.48	
TP-2	7.80	315.44	5.54		307.68	STONE
6+00				7.53	307.91	
+50				6.70	308.74	
7+00				5.92	309.52	
+50				5.19	310.25	
8+00				4.42	311.02	
+50				3.97	311.47	

+14.93
-14.49   ELEV. BM-23 = 315.00
+0.44   ELEV. HI = 315.44
          CHECKS → +0.44

J.M. Wilbourn

FIGURE 8-7. Profile leveling field notes include the station value and elevation at each station.

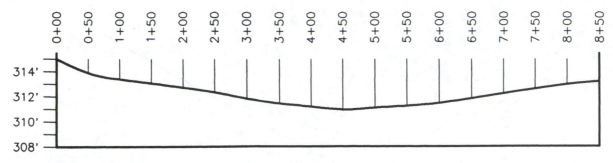

FIGURE 8–8. This profile drawing was created using the field notes shown in Figure 8–7.

Plotting Field Notes

Profile field notes can be plotted manually using a grid or profile paper, or using a CAD system with a similar prototype drawing containing a grid or series of horizontal and vertical guidelines. The bottom line of the profile represents the horizontal distance on the route of the survey and can be labeled with the station point values. The vertical aspect of the profile is an exaggerated scale labeled with elevation values. The elevation values of each station are taken from the field notes and plotted on the profile. An example is the profile created from the field notes in Figure 8–7 that is shown in Figure 8–8.

Profiles can be created for projects that consist of a single straight line, a series of connected straight lines (utilities), or straight lines connected by curves, such as highways. The profile information is plotted in the same manner, but the plan drawing or map indicates the true appearance of the feature. This is illustrated by the plan and profile drawing discussed in the next section.

PLAN AND PROFILE

Description and Uses

The plan and profile as used by civil engineers and state highway departments can be compared to a top and front view in mechanical drafting, where the front view is a section. This type of drawing is often much more detailed than the leveling profile, because it represents a completed project. The drawing is convenient to use, for it allows both the plan and cross section of a specific area to be shown on the same sheet. The plan is always placed above the profile. The uses of this type of map are many. Transportation departments use the plan and profile extensively for layout and design of roads and transit systems as seen in Figure 8–9. The profile view is often done along the centerline of a road to illustrate gradient and curves. Civil engineering firms employ the plan and

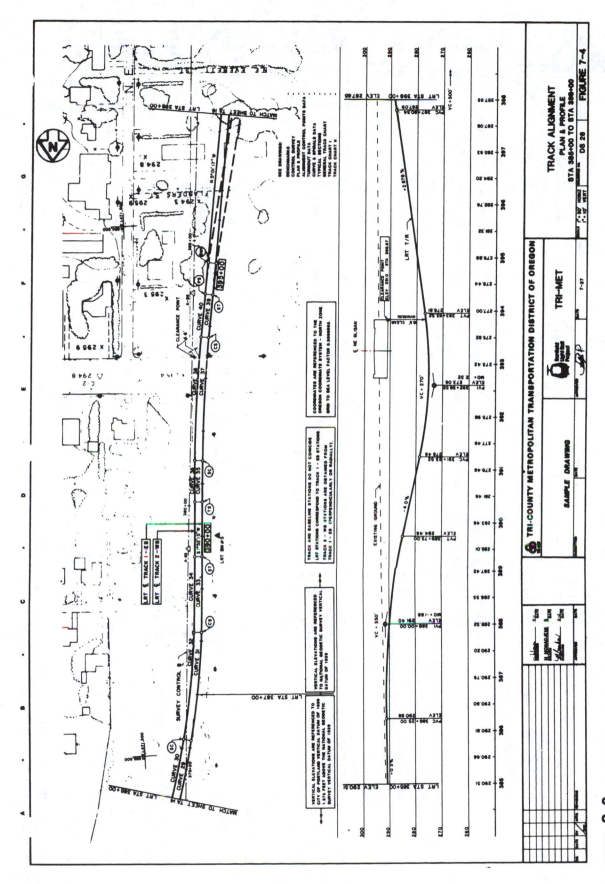

FIGURE 8–9. Plan and profile used in highway construction. *(Courtesy Oregon State Dept. of Transportation)*

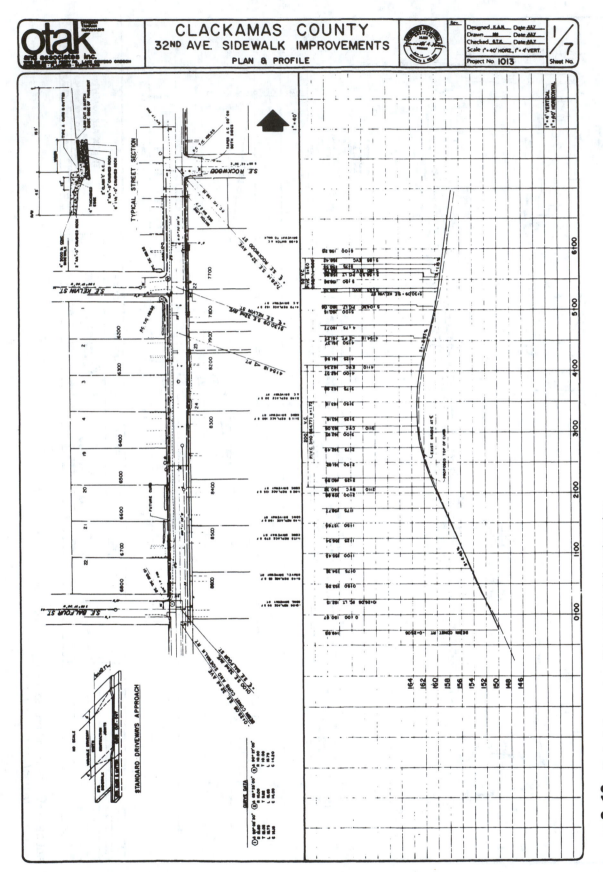

FIGURE 8–10. Typical plan and profile used by civil engineering companies. *(Courtesy OTAK & Associates, Inc.)*

profile when designing subdivision street layouts and underground utility locations (see Figure 8–10).

Layout and Construction

The scales most commonly used for the plan view are 1 in. = 100 ft and 1 in. = 50 ft. All pertinent information necessary for the map is drawn on the plan. The plan is normally long and narrow. because it is illustrating linear features, such as roads and sewer lines. Refer to Figure 8–10 for proper layout of the plan and profile. Before locating the plan on your drawing, it is important to know how much vertical elevation is to be shown in the profile.

The vertical scale of the profile is normally a ratio of 10:1 or 1 in. = 10 ft when the plan scale is 1 in. = 100 ft, and 1 in. = 5 ft when the plan scale is 1 in. = 50 ft. Using the appropriate scale, determine from the field notes the amount of elevation to be shown in the profile. This information allows the drafter to decide what size of paper or CAD drawing limits to use and where to locate the views. Because the profile is projected from the plan, the horizontal scale of the two views are the same.

Remember to allow some space beyond the highest and lowest elevations of the profile. Construct a grid of horizontal lines at even elevation points and vertical lines at even station points. Using this grid, the drafter can accurately locate manhole stations, grade, and sewer pipe elevations.

Terms and Symbols

The drafter should keep in mind that certain standard symbols and terms exist that are used frequently throughout the industry. But standards are sometimes modified, and each company may alter things to suit its needs. With this in mind, the drafter should always be aware of the company standards in use at the time.

Full station points are established by surveyors every 100 ft. The first station point is 00 + 00, then 1 + 00, 2 + 00, and so on. The example in Figure 8–10 shows a profile divided every 100 ft by vertical grid lines. At intervals along the plan view, manholes are located and their specific location is given as a station point value. The manhole symbol and station point value is shown in Figure 8–11 as they appear in both plan and profile. Manholes are abbreviated M.H., followed by an assigned number.

An important number called the *invert elevation* is always found on the profile view and is abbreviated I.E. This number represents the bottom inside of the pipe and is shown in Figure 8–11. This value is established in the design layout of the pipe or sewer line and is important in the surveying, excavating, and construc-

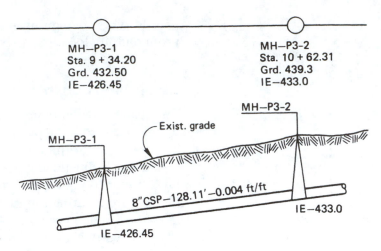

FIGURE **8–11.** Profile terms and symbols.

tion aspects of the job. Make it a habit to check the invert elevations on your drawing with either the engineer's sketch or the surveyor's field notes.

The distance between station points and the amount of slope are often indicated in the profile just above the pipe (see Figure 8–11). The size of the pipe is given first, then the distance between manholes followed by the slope, in this case vertical drop in feet to horizontal run per foot.

Grade is a common term that means an established elevation such as road grade. This is given with the manhole number and station value in the plan view. The drafter can use this number to plot grade elevation in the profile view.

Not all profiles are located directly under the plan, although this is the ideal situation. Some pipelines or sewers may take several turns through a new subdivision and the plan itself is not a linear shape. In this case the plan may be located to the left of the drawing and the profile on the right. The profile always appears as a straight line or flat plane, but the plan view may show several turns in the pipe. The profile is constructed in the same manner as we discussed, but the drafter is not able to project points from one view to the next.

PLAN AND PROFILE WITH CAD

Plan and profile sheets can be drawn with a CAD system in the same way they are manually drawn. This also applies to creating profiles of curved lines in which the curved line must be established along a related straight line. Or the plans and profiles can be generated automatically from contours and alignments. Some CAD packages will complete several procedures in one step, such as aligning the drawing to the sheet and trimming all parts outside the plan border. Profiles can also be edited. For example, the vertical scale can be changed while in the midst of a drawing.

There are also CAD packages that deal with sewer pipe design, drawing the manholes, and pipe cross sections. Refer to Figure 6–15 for an example of a plan and profile sheet of a sewer system drawn using a CAD program.

TEST

8–1   What is a map profile?

8–2   What are its uses?

8–3   Why is the vertical scale exaggerated?

8–4   Which points are projected from the map to the profile?

8–5   Define profile leveling.

8–6   How is the location of an IFS determined?

8–7   Write the following values in the proper station format.
300 –_____
567 – _____
298.43 – _____

8–8   What type of feature would require a profile leveling of connected straight lines and curves?

8–9   Who uses the plan and profile and for what purpose?

8–10   What scales are common for the plan and profile?

8–11   A point established every 100 ft in a linear survey is called a _____.

8–12   What is invert elevation?

8–13   What information may be found next to a manhole symbol on a plan and profile?

8–14   What is an established elevation called? _____.

PROBLEMS

P8–1 Construct a map profile along line *AB* in the map shown in Figure P7–3, in the last chapter. The vertical scale should be exaggerated. Use an A- or B-size sheet of vellum. Label vertical elevations and indicate scales. Use a proper sectioning symbol to indicate earth. Make a diazo print and submit to instructor for evaluation unless otherwise indicated.

P8–2 a. This problem requires you to plot a profile of a curved road. Construct your profile in the space provided on Figure P8–2 or on a separate sheet of vellum. Use the steps discussed in this chapter.

  b. Using your CAD system, digitize the plan view of Figure P8–2. Plot the profile of the curved road using the steps discussed in this chapter. Plot at a horizontal scale of 1 in. = 200 ft and a vertical scale of 1 in. = 20 ft on a C-size paper, and submit it to the instructor for evaluation.

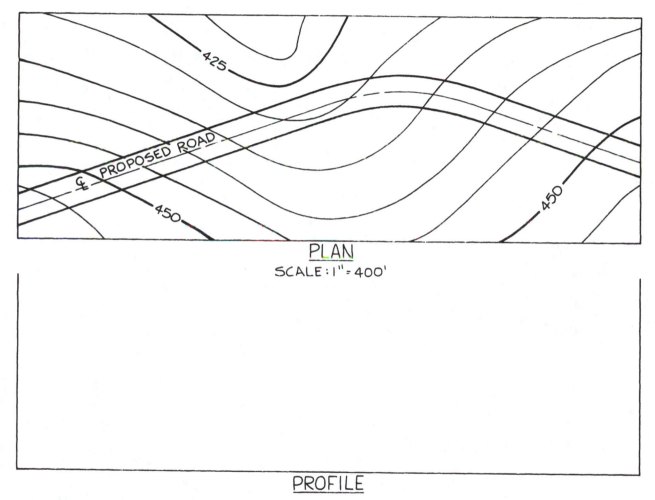

PLAN
SCALE: 1" = 400'

PROFILE

FIGURE P8–2.

P8–3 Survey stakes have been located along the centerline of a proposed street. A shot from the instrument to B.M.-1 has established an H.I. of 567.28 ft. Beginning at station 0 + 00, IFS shots have been taken at full stations for the length of the street, and their values are: 0.8, 2.6, 3.1, 3.8, 4.5, 5.1, 6.4, 7.1, 7.5, 8.3, 7.7, 7.5, 7.0, 6.2, 5.4, 9.6, 8.1, 4.7, 3.9, 3.5. Plot the profile for these stations. The final drawing should be plotted on a C-size piece of paper. Select the appropriate scales, and label the station and elevation values.

P8–4 a. Figure P8–4 is the plan view of a road and underground utilities. Using the information given, construct a plan and profile on a separate sheet of C-size vellum. Show all necessary information discussed in the text. Distances and slope between manholes should be labeled above the sewer pipe. Show water and gas lines in the plan view only. Make a diazo print and submit to instructor for evaluation unless otherwise indicated.

b. Using your CAD system and the information given in Figure P8–4, construct a plan and profile on a separate sheet of C-size paper. Show all necessary information discussed in the text. Distances and slopes between manholes should be labeled above the sewer pipe. Show water and gas lines in the plan view only. Plot at 1″ = 50′ and submit to instructor for evaluation.

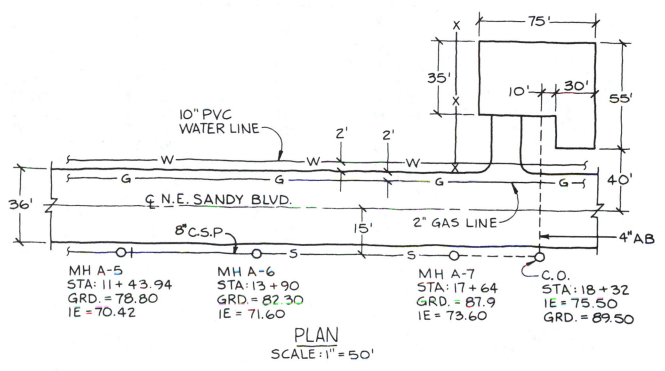

PLAN
SCALE: 1″ = 50′

FIGURE **P8–4.**

P8–5 Figure P8–5 is the plan view of a survey line of 1500 feet. Using a C-size vellum, draw the survey in scale of 1 in. = 50 ft and plot all of the existing features. Make proper note of the curve data and the points where the curve begins and ends. Show all data for location and the beginning transit line bearing. Show the north arrow with the scale of the drawing below.

Draw the survey line on the top half of the C-sheet, and record the survey line elevations on a grid on the lower half of the sheet. The grid should reflect the vertical elevation changes and the horizontal station locations.

Survey Line Elevations

Sta.	Elev.	Sta.	Elev.
15 + 00	119.0	22 + 60	124.0
15 + 30	120.0	23 + 00	123.0
15 + 70	120.5	23 + 50	121.0
16 + 00	120.0	23 + 75	120.0
16 + 60	118.0	24 + 00	119.0
17 + 00	117.0	25 + 00	117.75
18 + 00	119.0	25 + 50	117.25
18 + 70	120.0	26 + 00	117.75
19 + 00	120.5	26 + 50	118.0
19 + 60	121.0	27 + 00	119.0
20 + 00	121.5	27 + 50	119.0
20 + 50	122.0	28 + 00	118.0
20 + 80	124.0	29 + 00	114.0
21 + 00	125.0	29 + 50	112.0
22 + 00	126.0	30 + 00	110.0

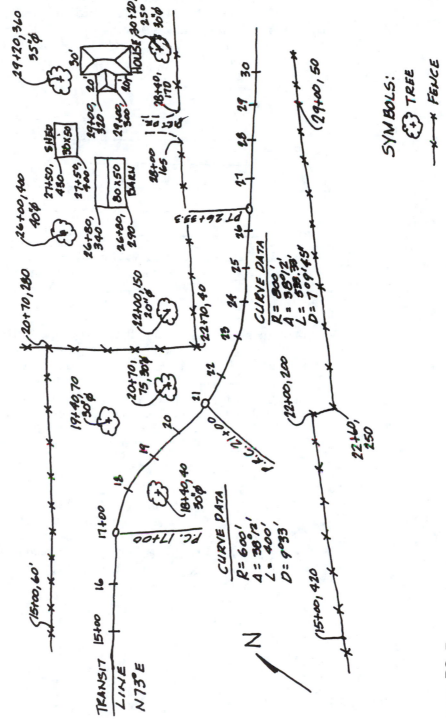

SURVEY FIELD NOTES

CURVE DATA
R= 600'
Δ= 38°12'
L= 400'
D= 9°33'

CURVE DATA
R= 800'
Δ= 38°12'
L= 533.33'
D= 7°09'45"

SYMBOLS:
⊕ TREE
×—×— FENCE

P8–6   Construct a set of six profiles across the piece of property shown in last chapter's Figure P7–8. One profile cuts through the transit station point A. The other profiles are 50 ft apart. Show all profiles on a single sheet of paper. Horizontal scale should be 1 in. = 100 ft, and the vertical scale should be 10 times the horizontal. Each profile should be only as long as the distance between property lines along the profile. Show elevation values for each profile. Label each profile with a subtitle, such as PROFILE A.

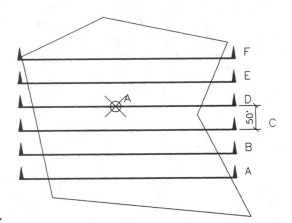

FIGURE **P8–6.**

P8–7   Construct a series of profiles across the stream in last chapter's P7–7. The first profile should be 50 ft from the southern property line, along the stream, and should be labeled PROFILE A. Label each profile consecutively. Each profile should be cut perpendicular to the stream, and extend for 100 ft beyond the stream, or to the edge of the property. Profiles should be 25 ft apart, and cover the entire length of the stream on the property. Determine the best horizontal and vertical scales to use in order to arrange all profiles on a D-size sheet of paper. Use a 10:1 ratio for horizontal and vertical scales. Label elevations on all profiles.

P8–8   Cut a profile along the stream in the drawing created in the previous problem, P8–7. Determine the best location for a small dam and pond with a water level that keeps the pond inside the property line. Construct any needed profiles to show the height of the dam and the water line. Add the dam and pond to the map. The drawings submitted for this problem should include the map and profiles, and should provide enough information to fully describe the size and location of the dam and pond.

P8–9   A sewer line is to be added to the house in P7–7. Run the sewer 30 ft from the west edge of house. The new sewer connects with a main sewer 30 ft south of the southern property line. Keep the new sewer 10 ft east of the driveway. Show a cleanout (CO) 10 ft from the house. Show an additional CO between house and main connection, no farther than 100 ft from the nearest CO. Use the following information for this problem:

> House sewer pipe is 4-in. ABS.
> Sewer main is 8-in. CSP.
> Grade at house is 608 ft.
> Sewer line IE at house is 603 ft.
> IE at house CO is 602.5 ft.
> Grade above sewer main is 584 ft.
> Sewer IE at connection point is 574.80 ft.

Create a plan and profile for the sewer installation. Do not show all the property. Show contours on the plan. Construct the drawing similar to the one in P8–4. In the profile along the sewer, note the type of pipe, distance between cleanouts, and slope in FT/FT. Choose appropriate scales and size of paper for this problem.

CHAPTER 9

Highway Layout

The layout of a proposed highway or road usually begins on a contour map or aerial photograph. Road designers and engineers, with input from government officials, determine the location of the road using maps and information gathered from field study. The engineer's initial design and centerline location is given to the surveyors, who then locate the proposed road's centerline and right-of-way boundaries. The surveyor's field notes are then plotted by the drafter on a contour map. When the initial layout is completed, the construction details can begin. This chapter discusses the initial layout of roads and highways.

The topics covered include:

- Plan layout
- Plan layout with a CAD system
- Profile layout
- Drawing conversions for CAD

PLAN LAYOUT

Centerline (Route) Survey

Before the drafter can begin the actual drawing of the road, a survey crew must physically locate the centerline on the ground and record bearings, distances, and station points as field notes. These notes are used by the drafter to plot the highway. The initial function of the survey crew is to mark the centerline of the highway. Subsequent surveys establish right-of-way, actual road widths, cross sections, and other details.

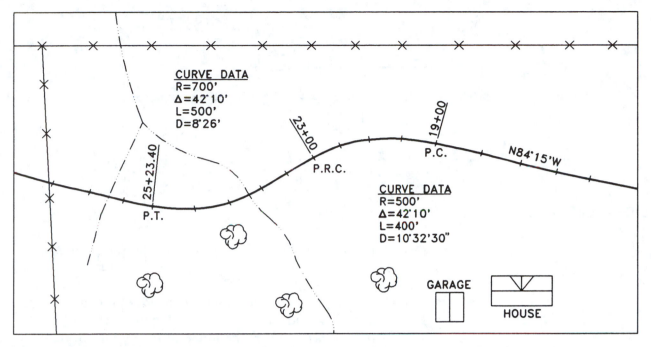

FIGURE 9-1. Plan of highway layout data.

CURVE DATA
R=700'
Δ=42°10'
L=500'
D=8°26'

25+23.40

23+00

P.R.C.

19+00

P.C.

N84°15'W

P.T.

CURVE DATA
R=500'
Δ=42°10'
L=400'
D=10°32'30"

GARAGE

HOUSE

Plan Layout

The plan and profile is an important drawing in highway layout. Using the surveyor's field notes, the drafter first constructs the plan view. This can be done on an existing map, or the drafter may have to draw a new one. The plan view shows trees, fences, buildings, other roads, and cultivated areas. The centerline, or *transit line*, is then drawn using bearings and distances. Station points are located every 100 ft. Figure 9-1 shows a plan layout. Notice that the road is to have curves. The information needed to construct the curve is given under the heading "Curve Data." Let us examine this information.

Point of curve (P.C.) is the point at which the curve begins. The station value is also written at the point of curve.

Radius (R) is a curve radius. To find the center point of this radius, you must project a line perpendicular to the centerline and measure the required distance on this line (see Figure 9-2).

Delta angle (Δ) is the central or included angle of the curve between the point of curve (P.C.) and the point of tangency (P.T.).

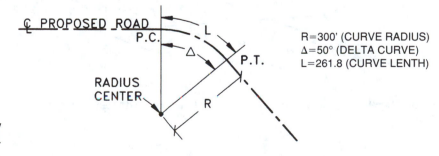

Ç PROPOSED ROAD
P.C.
L
Δ
P.T.
RADIUS CENTER
R

R=300' (CURVE RADIUS)
Δ=50° (DELTA CURVE)
L=261.8 (CURVE LENTH)

FIGURE 9-2. Basic curve data for highway layout.

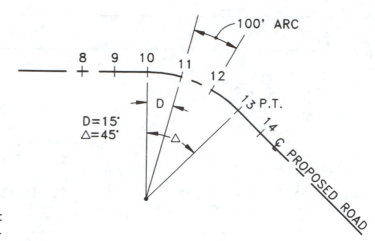

FIGURE 9–3. Degree of curve on a 100-ft chord.

From the perpendicular line projected to find the radius center, measure the delta angle. This point is the end of the curve (see Figure 9–2).

Curve length (L) is the centerline length of the arc from the P.C. to the P.T. or end of the curve (see Figure 9–2).

Degree of curve (D) is the angle of the 100-ft arc that connects station points. The degree is measured from the previous station (see Figure 9–3).

Point of reverse curve (P.R.C.) is the point at which one curve ends and the next curve begins. A *reverse curve* is known as an "S" curve in racing terminology and contains no straight section between curves. The station value is given after "P.R.C." (see Figure 9–1).

Point of tangency (P.T.) is the end of the curve. The station value is also written at this point.

Curve Layout: A Second Method

The survey crew may only establish centerlines and *points of intersection* (P.I.). A point of intersection on a plan view is the junction of two centerlines of different bearings. Curves are constructed at P.I.s. Figure 9–4 shows the curve layout method using bearings and P.I.s. The drafter is given bearings and distances and also suggested curve radii, and using this information must plot the road centerline and curves.

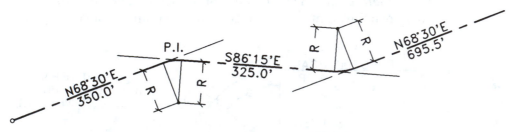

FIGURE 9–4. Highway curve layout using bearings, distances, and curve radii.

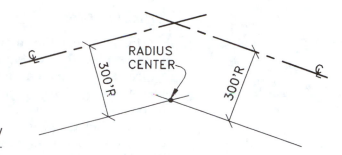

FIGURE **9–5.** Method of locating highway curve radius center.

To find the center point of the radius, measure perpendicularly from each leg of the road to the inside of the curve, the exact radius distance, as shown in Figure 9–5. Then draw a line parallel to each leg through the radius point just measured. The intersection of these two parallel lines is the center point for the radius curve. You can now construct the curve and road outline using the radius point.

Once the centerlines of the road have been plotted and the curves are laid out, the width of the road and right-of-way can be easily drawn by measuring from the centerline.

PLAN LAYOUT WITH A CAD SYSTEM

Constructing a plan view of a highway layout can be done with a CAD system in much the same way as with manual drafting. Data in the form of curve lengths, bearings, and distances can be entered when drawing the transit line. Some CAD software packages will routinely put in the centerline stationing.

The survey crew may establish centerlines and P.I.s. The CAD drafter then converts the curves to the proper radius. If the survey points are set relative to State Plane Coordinates, the CAD drafter can overlay this on a digital topographic map. This will automatically place the highway layout on the existing surface, which will be helpful in constructing the profile.

PROFILE LAYOUT

The plan view shows all necessary horizontal control: bearings, distances, radii, and angles. But as in the location and construction of a sewer line, a road requires vertical control. The ground does not remain flat, and the road must reflect this. All necessary vertical information is shown on the profile. The layout of the profile is done in the manner discussed in Chapter 8.

Construct a grid for the profile and label vertical elevations along the side. If there are contour lines on the plan, the labels should coincide with them. Along the bottom of the profile, the stations should be labeled (every 100 ft). Next, the profile of the present ground level should be plotted and drawn.

Vertical Curves

A vertical curve is the shape of the road or highway as it crests a hill or reaches the bottom of a valley and creates a "sag." These features are calculated mathematically by the engineer or computer program. The drafter is given all necessary elevation points and then plots the curve. In Figure 9–6, the B.V.C. (begin vertical curve) is labeled, as is the E.V.C. (end vertical curve). The vertical curve occupies this entire distance and is labeled on the profile as 200 ft. V.C.

Keep in mind that vertical curves are measured as a horizontal distance and not a radius.

The P.I. (point of intersection) on the profile is the intersection of the projected grade lines or grade slopes. The elevation of this point can be calculated using the elevations of the B.V.C. and E.V.C. and grade slopes. Remember that grades are given as percentages.

The vertical curve is tangent to two points, stations 4 + 00 and 6 + 00. A straight line connecting these two stations has an elevation of 761.55 ft directly below the P.I., and the vertical distance between the P.I. and the straight line is 5.00 ft. The vertical curve will pass midway through this distance and have an elevation of 764.05 ft. The distances from the grade line to the curve at each station point can be calculated using the formula

$$\frac{(D_1)^2 h}{(D)^2} = TD$$

FIGURE 9–6. Vertical curve layout.

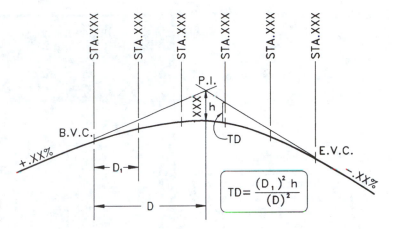

FIGURE **9-7.** Components of vertical curve formula.

$$TD= \frac{(D_1)^2\, h}{(D)^2}$$

The tangent distance (TD) is the measurement from the grade line to the profile of the curve at a station point. Once you have solved the formula for this distance, it can be subtracted from the elevation of the grade line to give you the elevation of the road profile at a specific station point.

Figure 9–7 graphically identifies all the components of the formula. The distance between the B.V.C. and the P.I. is labeled D, and the distance from the B.V.C. to the required station is D_1. The height of the midway point directly below the P.I. is termed h. It can be stated that the distances from the grade line to the road profile are proportional to the squares of the horizontal distances from the tangent points.

Table 9–1 provides the calculations for all the points of the vertical curve in Figure 9–6. The *ordinate* is the tangent distance from the top of the curve to the grade (tangent) line at each horizontal station point.

The drafter can plot the points of the vertical curve at each station and then connect those points to create the curve. This line becomes the profile of the proposed road. Elevations of the road are written vertically at each station point.

TABLE **9-1.** Points of the vertical curve in Figure 9-6

Station	Tangent Elevations	Ordinate	Curve Elevations
4 + 00	761.05	0	761.05
4 + 25	762.43	0.16	762.27
4 + 50	763.81	0.63	763.18
4 + 75	765.18	1.41	763.77
5 + 00	766.55	2.50	764.05
5 + 25	765.43	1.41	764.02
5 + 50	764.31	0.63	763.68
5 + 75	763.18	0.16	763.02
6 + 00	762.05	0	762.05

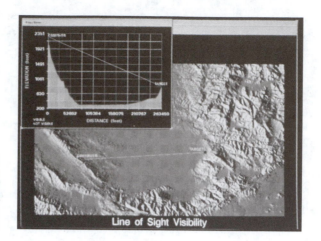

Line of Sight Visibility

FIGURE 9–8. Visibility studies bring engineering functions to planning and site location processes. This image was created with ARC/INFO® software. (Graphic image supplied courtesy of Environmental Systems Research Institute Inc.)

Figure 9–8 shows a visibility study done with a CAD system. These kinds of studies can be useful and efficient when used in conjunction with highway design. Besides using a visibility study, it is possible to use CAD to design highway layouts by including user-defined input such as stopping sight distance (the distance needed to stop safely), passing sight distance (the distance needed to pass safely), and vertical curve limitations.

DRAWING CONVERSIONS FOR CAD

CAD drawings can be created from existing paper-based drawings by the use of a *scanner*. A scanner is a device that the operator can use to "read" the lines on the existing drawing. The operator either drags the scanner over the drawing (as in the case of a hand-held scanner), or inserts the drawing into a more sophisticated scanner (which may resemble a type of blue-line machine). The images that the scanner reads and picks up are called *raster images*, and the resulting electronic file is called a *raster file*. A raster image is nothing more than a record of individual "dots" which are either on or off, depending on whether a line has been read by the scanner or the background has been read by the scanner.

CAD systems operate using *vectors*. A vector is a line which has a specific length, specific end points, and a specific direction. It differs from a raster in that a raster only *looks* like a vector. When a scanner creates a raster file, the file needs to be converted to a vector file using a conversion program. The vector file can then be manipulated by the CAD drafter.

Scanning a drawing and then converting the raster file to a vector file is much faster than digitizing the drawing. Scanning therefore has many applications. For example, a road 40 ft wide is to be built across a tract of land. If an existing topographic map drawn on paper is available, a scanner can be used to put that information into a CAD system. Then the design of the road can be created within the CAD system.

TEST

9–1 What type of survey is used to lay out the road initially? _____

_____

9–2 What is the point at which the curve begins? _____

_____

9–3 What is the delta angle? _____

_____

9–4 How is the degree of curve measured? Show your answer in a sketch.

_____

9–5 What is the point at the end of the curve called? _____

_____

9–6 Define the point of intersection (P.I.). _____

_____

9–7 Briefly describe how you would locate the center point of a 400-ft-radius horizontal curve, given two centerlines and a P.I.

_____

9–8 Why is a profile necessary for road construction? _____

_____

9–9 What is a vertical curve? _____

_____

9–10 What type of information does the vertical curve show? _____

_____

9–11 What is a scanner, and how does it apply to computer-aided drafting?

_____

_____

9–12 What is the difference between a raster drawing and a vector drawing?

_____

_____

9–13 What use does a Visibility Study have?

_____

_____

_____

PROBLEMS

P9–1 a. For this problem you will lay out a 40-ft-wide road using the information given in Figure P9-1. Plot the centerline bearings on Figure P9-1, then construct the curves and road outlines. Label all points, bearings, and radius curves.

b. If you have access to a scanner and a CAD system, scan the information given in Figure P9-1. If you have a CAD system only, then digitize the information given in Figure P9-1. Be sure that the contours and existing road have real world coordinates. Construct the centerline bearings on Figure P9-1 first, then construct the curves and road outlines. Label all points, bearings, and radius curves. Show centerline stationing along the road. Plot at a scale of 1 in. = 200 ft on an 8½ × 11-in. sheet of paper, and submit it to your instructor.

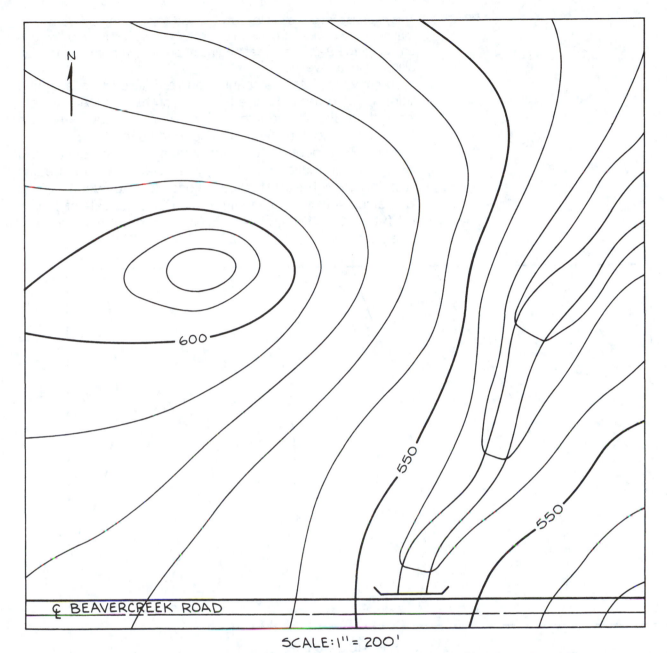

SCALE: 1" = 200'

Point A: 710' west from right edge of map on ℄ of Beavercreek Rd.
Point B: N24°30'E, 550', 200' radius curve
Point C: N39°25'W, 620', 300' radius curve, then due north to edge of map.
Road to be 40' wide.

FIGURE P9–1.

P9–2   a. This exercise involves the layout of a road using some different information. A 40-ft-wide street is to be plotted using the information given in Figure P9–2. The street begins at point A, which is 100 ft east along Holgate Avenue.

b. If you have access to a scanner and a CAD system, scan the information given in Figure P9–2. If you have a CAD system only, then digitize the information given in Figure P9–2. Be sure that the contours and existing road have real world coordinates. To repeat, a 40-ft-wide street is to be constructed using the information given in Figure P9–2. The street begins at point A, which is 100 feet east along Holgate Avenue. Show centerline stationing along the road. Plot at a scale of 1 in. = 50 ft on 11 × 17-in. sheet of paper, and submit it to your instructor.

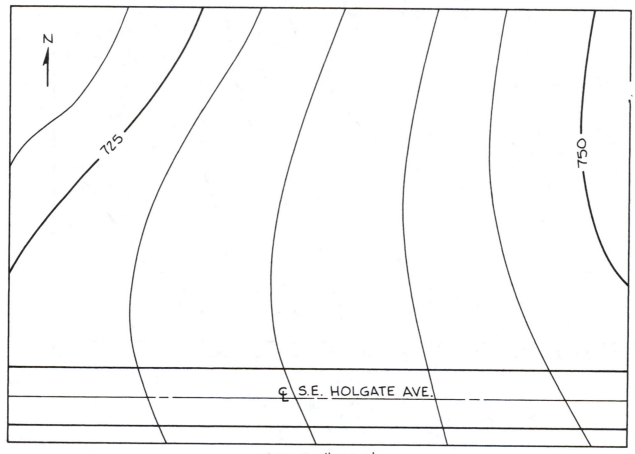

SCALE: 1" = 100'

Point A: 100' east from left edge of map on ℄ S.E. Holgate Ave.
Point B: Due north 110'
Point C: Δ angle – 30° (NE)
    Radius curve – 150' to point C

Point D: 170' from C
Point E: Δ angle – 60° (NE)
    Radius curve – 100' to point E
Point F: 160' from E

Point G: Δ angle – 90°
    Radius curve – 120' to point G
Point H: Due south to ℄ Holgate Ave.
Plot A 40' wide street.

FIGURE P9–2.

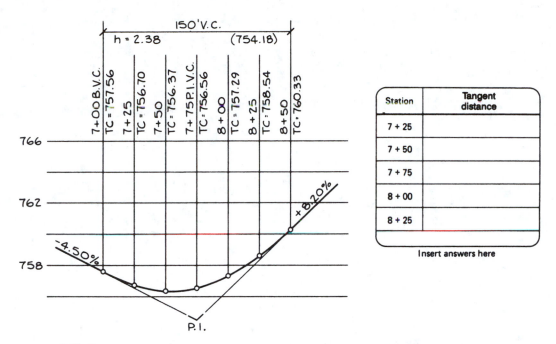

FIGURE P9–3.

Station	Tangent distance
7 + 25	
7 + 50	
7 + 75	
8 + 00	
8 + 25	

Insert answers here

P9–3 Figure P9–3 contains enough information for you to calculate the elevation of the road at each station point on the vertical curve. Using the formula given in this chapter, make your calculations and record them in the space provided.

CHAPTER 10

Earthwork

Earthwork is a term used in construction to describe quantities of earth either excavated from areas or filled into low spots. *Cut* and *fill* represents earth that is cut from hillsides and filled into valleys and low areas. *Borrow pits* are excavations along highways or construction projects from which material is removed and used as fill elsewhere. This chapter describes how earthwork can be plotted on maps for highways and site plans, and how the volumes of earth can be calculated with mathematical formulae.

Topics covered include:

- Establishing the cross-sectional profile
- Locating the cut-and-fill boundaries
- Plotting cuts and fills for an inclined road
- Highway cut-and-fill layout with a CAD system
- Site plan cut-and-fill layout
- Borrow pit calculations

HIGHWAY CUT-AND-FILL LAYOUT

Level Road

In designing or planning the location of a road, it is necessary to determine accurately how far the cuts and fills will extend beyond the sides of the road. This is important in determining the proper amount of land to be purchased for the right-of-way.

The road is first plotted on a contour map using surveying field data, and one of the highway layout techniques discussed in

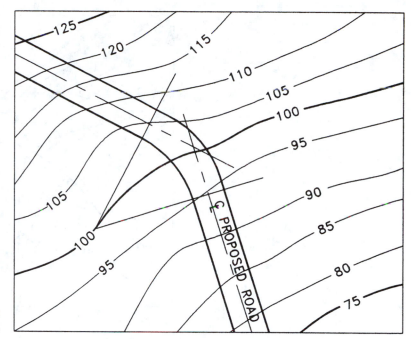

FIGURE 10–1a. Begin cut-and-fill layout by plotting the new road on the map.

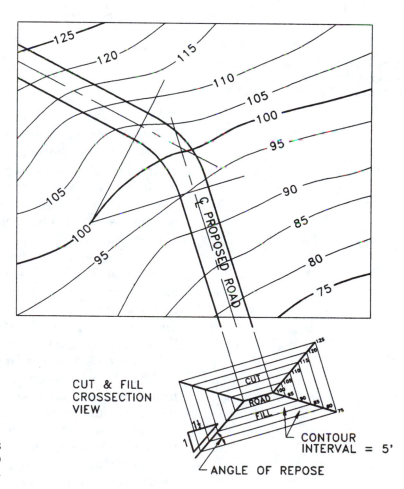

CUT & FILL
CROSSECTION
VIEW

CUT

ROAD

FILL

CONTOUR
INTERVAL = 5'

ANGLE OF REPOSE

FIGURE 10–1b. Cut-and-fill cross section is added to map perpendicular to road.

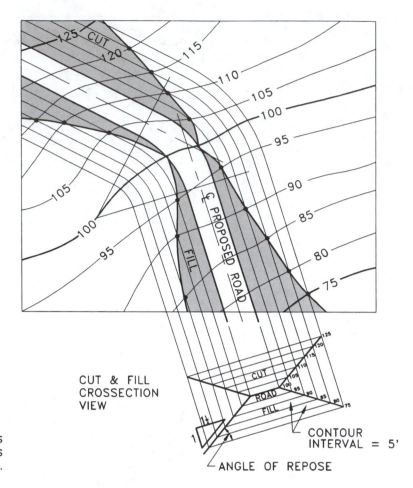

CUT & FILL
CROSSECTION
VIEW

CONTOUR
INTERVAL = 5'

ANGLE OF REPOSE

FIGURE 10–1C. Intersection of cross section values with map contour lines permits outlines of cut and fill to be drawn.

Chapter 9 (see Figure 10–1a). The engineer establishes an *angle of repose*, which is the slope of the cut and fill from the road. The angle of repose is basically the ratio of run to rise (Figure 10–2) and is determined primarily by the type of soil or rock to be cut through or used as fill material. An angle of repose of 2:1 means that for every 2 ft of horizontal distance, the slope rises or drops 1 ft vertically.

A cross-sectional view, or profile, is established off the end of the road. Then the angle of repose slope is plotted. Using an angle of repose of 1½:1 for an area of cut, the slope is determined by measuring from the edge of the road 1½ units horizontally. Then measure vertically one unit and mark that point. A straight line connecting the edge of the road and the point just measured reflects the 1½:1 angle of repose. See Figure 10–1b.

The location of contour lines on this slope is determined by measuring vertically from the road using the horizontal scale of the map. Mark each contour value above and below the road. Project these marks to the slope lines to find the contour locations. See Figure 10–1b.

The points at which the contour values intersect the angle of repose in the cross section are projected onto the map and paral-

FIGURE 10–2. Angle of repose is the same as the ratio of run to rise.

lel to the road. Project all lines as shown in Figure 10–1c. Cut and fill boundaries, such as the top of a cut and the toe of a fill, are located next. To plot the cut in Figure 10–1c, first follow the line projecting from the 105-ft contour in the cross section up into the map and make a mark where this line intersects the 105-ft contour line on the map. Do the same with the 110-, 115-, and 120-ft contour marks in the cross section. Always follow each line from the cross section completely through the map. The line may cross its contour more than once.

Use this same procedure to plot the cut on both sides of the road and to plot the fill. If cut and fill are the same angle of repose, as in Figure 10–1c, the lines projected from the cross section represent both cut and fill values. But if cut and fill are different angles, it is best to project the cut lines from the cross section only along areas of the map that are above the level of the road. In a like manner, project scale lines for fill from the cross section only into areas of the map that are below the level of the road. This prevents the confusion of having two sets of lines along the entire length of the road.

To complete the cut-and-fill boundaries, begin drawing a line at the intersection of the road and the contour line having the same elevation as the road. Keep the cut and fill boundary line

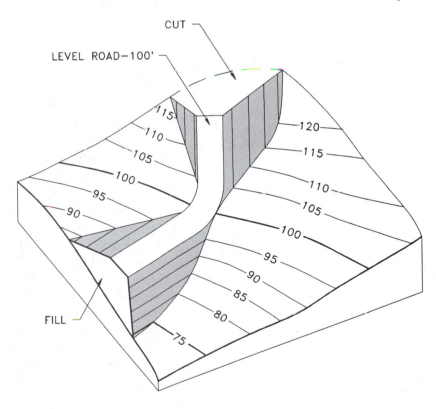

FIGURE 10–3. Pictorial view of cut and fill.

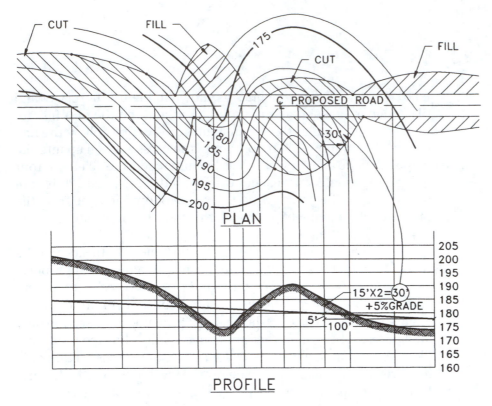

FIGURE 10–4. Cut-and-fill layout for an inclined road.

between the lines projected from the cross section until you reach the next mark on a contour line. See Figure 10–1c.

Figure 10–3 is a pictorial representation of the map in Figure 10–1.

Inclined Road

The amount of cut and fill for an inclined road can best be established with a plan and profile. The road is plotted on the contour map as for a level road, but a profile of the existing grade along the entire centerline is drawn as shown in Figure 10–4.

The *grade* of the road is determined and plotted on the profile. The grade is termed "percent of grade." A 1 percent grade rises 1 ft vertically for every 100 ft of horizontal distance. Therefore, a 100 percent grade is a 45° angle.

The road in Figure 10–4 has a 5 percent grade and an angle of repose of 2:1 for both cut and fill. To establish cut and fill, first project the intersections of each contour line and road centerline from the map to the profile drawing. At the point where the line projected from the road centerline on the map intersects the road grade in the profile, measure from the road to the surface of the natural ground line. For example, the intersection of contour 185 is projected from the road centerline in the map to the profile. The measurement along this line in the profile from the road to the nat-

---

ural ground elevation is 15 ft. Multiply this figure by the angle of repose (2) and place that measurement perpendicular to the edge of the road in the map plan. Project the point parallel to the road to the appropriate contour, as shown in Figure 10–4. Do this for all the contours that cross the road in the profile. The points can then be connected to delineate the areas of cut and fill.

HIGHWAY CUT-AND-FILL LAYOUT WITH A CAD SYSTEM

With a CAD system and three-dimensional survey information, a highway cut-and-fill design can be constructed with relative ease. The highway centerline, elevations, and grade are chosen along the digital topographic map. With the appropriate CAD software and with angle of repose input from the user, the cut-and-fill can be calculated and drawn automatically. The CAD system may also calculate the volumes of cuts and fills along the highway.

CROSS SECTIONS

The previous discussion of highway cut-and-fill layout covered a method used for plotting cut and fill using only route survey data of a road centerline. After laying out the highway, the angles of repose were used to plot the cut and fill.

The cross section method of plotting cut and fill is based on field surveys, or lines of levels run perpendicular to the road. These level lines are often short, and span only the width of the highway and areas affected by the cut or fill. Cross sections are important because they provide measurements that can be used for calculating the amounts of earth to be removed from hills and filled into valleys and areas below road grade.

Cross Section Surveys

Profiles perpendicular to the road centerline are usually measured at full stations (every 100 ft), at 50-ft stations, and at breaks, or changes in the profile of the center line. The sections extend far enough from the center line to include any cut or fill.

The level instrument is set up at a station that allows several profile measurements to be made from one instrument location. Each foresight reading along the center line is noted as an *intermediate foresight (IFS)*. The IFS is located at a station or elevation change. Additional rod readings are taken at set distances perpendicular to the centerline. For example, an IFS reading of 6.4 is recorded on the center line. Then the rod is moved 15 ft perpendicular from the centerline and a rod reading, or foresight (FS) of 7.9 is recorded. The elevation difference is -1.5 ft (6.4 − 7.9 = -1.5). A second FS of 9.2 is measured 35 ft perpendicular from the center line, and the elevation difference from the center line is

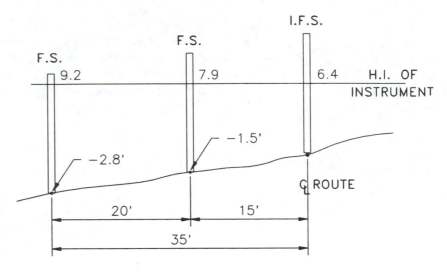

FIGURE 10–5. Cross section surveys are composed of several elevation measurements taken perpendicular to the center line.

−2.8 ft. (6.4 − 9.2 = −2.8). See Figure 10–5. This process is used on both sides of the center line.

Cross Section Field Notes

Field notes for level profiles and cross sections contain all the information required to construct cross section drawings. Figure 10–6 shows an example of cross section field notes. Surveyors try to measure several cross sections before moving to another instrument setup. A TP measurement is taken to relocate the instrument. Several IFS readings can then be taken from one sta-

CROSS-SECTIONS FOR ROAD 213

APRIL 22, 1993
SUNNY - 65°
LIETZ LEVEL #771

π – J.M. WILBOURN
ROD – B.A. DOTY
NOTES – M.R. CULVER

STA.	B.S.	H.I.	F.S.	I.F.S.	ELEV.	LEFT		₵	RIGHT	
7+00				4.21	877.13	$-\frac{0.8}{35}$	$+\frac{0.4}{15}$		$+\frac{0.5}{14}$	$+\frac{1.9}{35}$
6+50				4.37	876.97	$-\frac{0.2}{35}$	$+\frac{.5}{16}$		$-\frac{0.9}{15}$	$+\frac{0.4}{35}$
T.P.	6.22	881.34	4.67		875.12					
6+00				5.85	874.01	$-\frac{0.8}{35}$	$-\frac{0.3}{15}$		$+\frac{0.4}{16}$	$+\frac{0.2}{35}$
5+75				5.67	874.19	$-\frac{0.4}{35}$	$-\frac{0.6}{15}$		$+\frac{0.8}{15}$	$+\frac{1.7}{35}$
5+50				5.25	874.61	$-\frac{1.1}{35}$	$-\frac{0.2}{14}$		$+\frac{0.5}{15}$	$-\frac{0.7}{35}$
5+00				4.73	874.93	$+\frac{1.3}{35}$	$+\frac{0.6}{15}$		$-\frac{0.2}{15}$	$+\frac{0.1}{35}$
B.M.-12	3.62	879.86	4.74		876.24					

FIGURE 10–6. Cross section field notes contain all the information required to construct cross section drawings.

tion. Surveyors may also begin their cross section notes at the bottom of the page. This places measurements on the right side of the center line, on the right side of the page.

The elevation notes for the profile are usually given on the left side of the page, and the measurements for the cross section appear on the right. Notice that the cross section notes contain what appear to be positive and negative fractions. The numerator is the elevation difference from the centerline IFS. The denominator represents the distance of that point from the center line. The plus or minus indicates whether the point is above or below the elevation of the IFS on the center line.

SITE PLAN CUT-AND-FILL LAYOUT

The following discussion outlines the procedure that can be used to plot the cut and fill surrounding a site without using surveyor's profile or cross section notes. This procedure can be used to estimate the approximate limits of cut and fill. A series of profile cross sections can be surveyed at the site to determine more exact areas of cut and fill.

Site Plan Layout

The property boundaries for the proposed site are plotted on a contour map using surveyor's notes. The elevation of the site

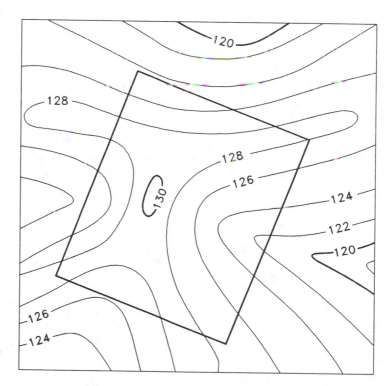

FIGURE 10–7. Site plan layout before application of cut and fill.

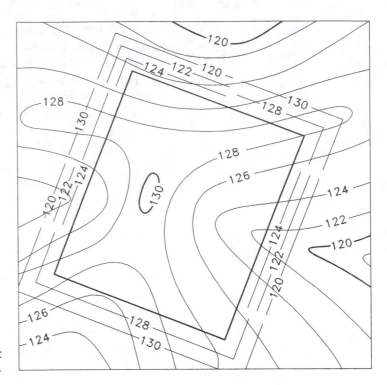

FIGURE 10–8. Site plan layout with cut and fill scales added.

(after excavation or filling) is determined by the engineer, as is the required angle of repose. This information may be given on the plan as well as bearings and distances for the property lines. Figure 10–7 depicts the site plan layout before application of cut and fill.

Cut and Fill

The drafter should first determine the areas of cut and fill. Then the appropriate scales for each can be plotted on the map. In Figure 10–8, the angle of repose for cuts is 2:1 and 1½:1 for fills. The contour interval is 2 ft. In areas of fill, measure perpendicular to the property lines 1½ times the contour interval and draw a line parallel to the property line. For the areas of cut, measure perpendicular to the property lines two times the contour interval, or 4 ft, and draw lines parallel to the site boundaries. Study the example in Figure 10–8 closely.

The elevation for the site is to be 126 ft. Areas below this elevation are fill and areas above are cut. Where the first line of the fill scale intersects the 124-ft contour, make a mark. The second line of the fill scale is the 122-ft contour. Find all the intersections of like contours and then connect those points as shown in Figure 10–9. Use the same technique when finding the required points for the cut. The first line parallel to the property line on the cut scale is 128 ft.

From this map the amounts of excavation and fill can be estimated. Any profiles that are needed can be taken directly from the

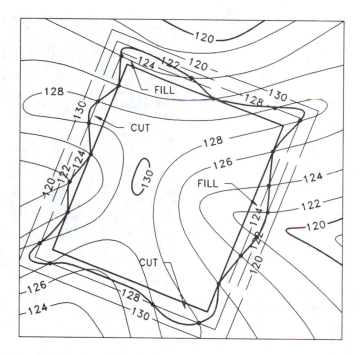

FIGURE **10–9.** Completed site plan layout showing areas of cut and fill.

map. Legends and additional information can be placed on the map as required.

SITE PLAN CUT-AND-FILL LAYOUT WITH A CAD SYSTEM

A CAD system gives the drafter the opportunity to view site plan layouts graphically. A three-dimensional survey or digital topographic map is used with a CAD software program. This allows the drafter the opportunity to enter site slope information and

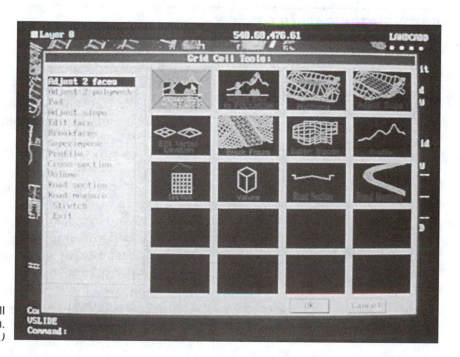

FIGURE **10–10.** Site plan cut-and-fill details possible with a CAD system. *(Courtesy of LANDCADD International, Inc.)*

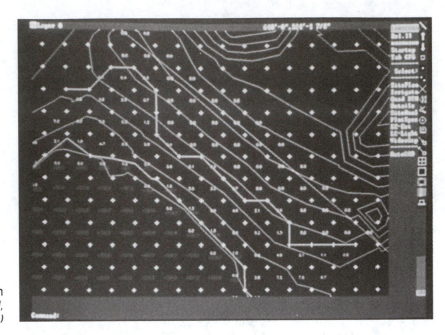

FIGURE 10–11. A site plan drawn with CAD. *(Courtesy of LANDCADD International, Inc.)*

angle of repose to construct the design site plan. Figure 10–10 shows some examples of the type of details that are possible to modify when working with a CAD system. Adjustments can then be made to the site plan design with relatively little effort. A site plan such as the one shown in Figure 10–11 can be generated from existing topography and the design criteria. The site plan can also show the amount of cut or fill necessary within each grid. The CAD drafter can also produce any necessary profiles from the established plan data.

EARTHWORK CALCULATIONS

Borrow Pits

A *borrow pit* is an area from which quantities of earth are excavated to construct embankments and fills. Borrow pits are often visible at interchanges along interstate highways. They can appear as square, rectangular, or trapezoid ponds, because they often fill with water.

Accurate calculations of borrow material are important because the contractor is paid based on the amount of earth removed. The price per cubic yard is multiplied by the number of yards excavated and hauled.

The borrow pit is defined by laying out a grid. The size of the grid is determined by the amount of material needed, the type of soil, and the land that is available. The grid is composed of squares with sides of 20, 40, 50 or 100 ft. See Figure 10–12. The elevations of each grid intersection are determined by running levels. Reference points, or hubs, are established on extension lines beyond the limits of the grid. Using these reference hubs, the grid can be easily relocated after the excavation.

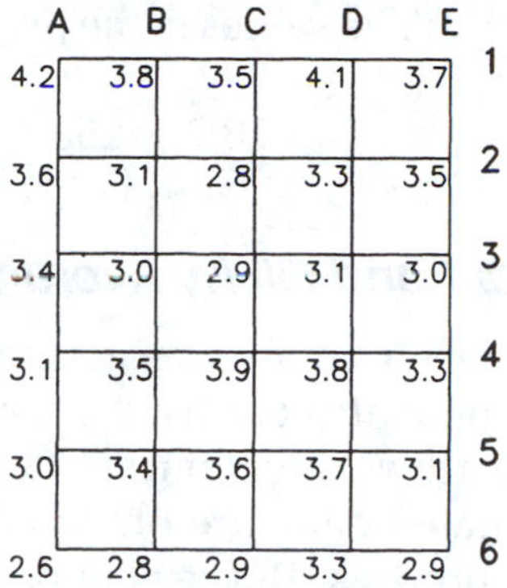

FIGURE 10-12. The borrow pit is laid out as a grid, and the final excavation depths are labeled at each grid intersection.

The elevations of the grid intersections are determined after excavation by running levels again. The heights of the cut are recorded for each intersection, and are shown in Figure 10-12. These values can then be used to determine the quantity of borrow material. The quantity of earthwork in one of the squares is calculated by multiplying the area of the section by the average of cut depths at each corner. For example, in Figure 10-12, the square A1, A2, B1, B2 is 50 ft × 50 ft, or 2500 sq ft. The average depth of its four corners is:

$$\frac{4.2 + 3.8 + 3.6 + 3.1}{4} = 3.675$$

Multiply 2500 × 3.675 and the result is 9187.5 cu ft. Divide cubic feet by 27 to obtain cubic yards:

$$\frac{9187.5}{27} = 340.28 \text{ cu yds}$$

The most accurate calculations can be obtained by following this procedure for each square in the grid. The procedure can be simplified by combining squares of the same dimensions. For example, to find the material in the larger square of A1, A3, C1, C3, you must find the average of all the corner depths for four squares. Corners that are common to more than one square must be added for each square. The total of the corner depths for the large square just defined is 52.8. Divide this by 4 and the average cut depth is 13.2 ft. Finally, the average cut depth multiplied by the area of the square (2500 ft) is 33,000 cu ft, or 1222 cu yds. One cubic yard is equal to 27 cubic feet.

The formula for determining the amount of material is:

Area of section × Average of corner depths = Cubic feet

Cubic yards is then calculated as follows:

$$\frac{\text{Cubic feet}}{27} = \text{Cubic yards}$$

Earthwork

245

The values in the previous example are calculated as follows:

$$13.2 \times 2500 = 33,000 \div 27 = 1222 \text{ cu yd}$$

Cut and Fill by Average End Method

Maps that are constructed using CAD programs and specialized application software can contain enough data for calculating earthwork volumes. These procedures, based on computer-generated data, are often sufficient for construction and estimating purposes. But it is important for surveyors and civil drafters to understand the nature of cross section area calculations and to be able to calculate volumes of earthwork in cut and fill.

A *level* cross section is one in which the ground is basically level and the height, or depth, of the cut is constant across the profile. The area (A) of a level cross section is calculated by figuring the average of the top and bottom widths (d and w), and multiplying this by the height (h) of the cut (or fill). See Figure 10–13.

$$A = h\left(\frac{2d + w}{2}\right)$$

$$A = h\left(d + \frac{w}{2}\right)$$

Cross sections that are not level, but have three distinct levels, are calculated by finding the area of triangles. See Figure 10–14. The following formula can be used for these calculations:

$$A = \left(\frac{1}{2}\right)\left(\frac{w}{2}\right)\left(h_1 + h_2\right) + \left(\frac{1}{2}\right)\left(c\right)\left(d_1 + d_2\right)$$

$$A = \frac{w}{4}\left(h_1 + h_2\right) + \frac{c}{2}\left(d_1 + d_2\right)$$

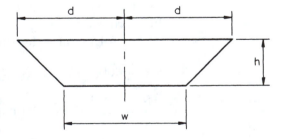

FIGURE 10–13. The area of a level cross section is calculated by using the dimensions of the profile.

These two formulae allow you to calculate the area of a cross section. These areas can then be used to calculate earthwork volumes. The *average end method* uses the averages of the two end cross sections to compute the volume of earth between the sections. The volume (V) is equal to the average area of the two cross sections (A_1 and A_2), multiplied by the distance between the two (D). Cubic yards is determined by dividing the result by 27.

$$V = \left(\frac{A_1 + A_2}{2}\right)\left(\frac{D}{27}\right)$$

The following example illustrates how averages can be calculated for the two cross sections shown in Figure 10–15. The averages are then used to calculate the earthwork volumes between the two sections.

Section 8+00

$$A = \left(\frac{1}{2}\right)\left(\frac{w}{2}\right)\left(h_1 + h_2\right) + \left(\frac{1}{2}\right)(c)\left(d_1 + d_2\right)$$

$$A = \frac{w}{4}\left(h_1 + h_2\right) + \frac{c}{2}\left(d_1 + d_2\right)$$

$$A = \frac{15}{4}\left(15 + 12\right) + \frac{8}{2}\left(32 + 27\right)$$

$$A = 101.25 + 236 = 337.25 \text{ sq ft}$$

Section 9+00

$$A = \frac{w}{4}\left(h_1 + h_2\right) + \frac{c}{2}\left(d_1 + d_2\right)$$

$$A = \frac{15}{4}\left(13 + 9\right) + \frac{8}{2}\left(28 + 24\right)$$

$$A = 82.5 + 208 = 290.5 \text{ sq ft}$$

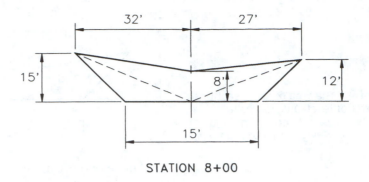

STATION 8+00

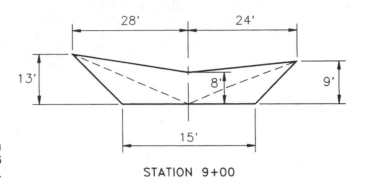

FIGURE **10–15.** The volume of earth between these two cross sections is 1161.36 cu yd.

STATION 9+00

Now that the areas of the two sections have been determined, an earthwork calculation for the volume of material between the two sections can be made. The following formula allows you to find the volume in cubic yards. In this formula, A equals the area of a cross section, and D is the distance between cross sections.

$$V = \left(\frac{A_1 + A_2}{2}\right)\left(\frac{D}{27}\right)$$

$$V = \left(\frac{337.25 + 290.5}{2}\right)\left(\frac{100}{27}\right)$$

$$V = 3.3.88 \times 3.70 = 1161.36 \text{ cu yds}$$

TEST

10–1 What is cut and fill?_____

_____

10–2 Why is it necessary that cuts and fills be accurate?_____

_____

10–3 What is the slope of a cut determined by?_____

_____

10–4 What is the slope of a cut and fill called?_____

_____

10–5 What is percent of grade?_____

_____

10–6 Why might industrial sites require the calculation of cuts and fills?

_____

_____

10–7 What is a cross section?_____

_____

10–8 What is an intermediate foresight?_____

_____

10–9 How many IFS readings can be taken from a single instrument setup?

_____

_____

10–10 Write the correct field notes notation for an IFS reading that is 25 ft from the center line and 2.6 ft below the elevation of the center line.

_____

_____

10–11   What determines the size of a borrow pit?_____

_____

_____

10–12   What is the meaning of the numbers recorded at the intersections of the borrow pit grid?_____

_____

10–13   Write the formula for obtaining the cubic feet in a borrow pit, and the formula for converting cubic feet into cubic yards._____

_____

10–14   Define the *average end method* of computing earth volumes.

_____

_____

PROBLEMS

P10–1   a. Figure P10–1 shows a proposed road layout. It will be a level road at an elevation of 270 ft. Angle of repose for cut and fill is 2:1. Construct the required cut and fill directly on the map. Your cut-and-fill scale may be placed in the space provided. Show all of your work. Use different shading techniques for cut and fill.

b. Transfer the graphics shown in Figure P10–1 to a CAD system, either by digitizing or by scanning. Using the information given in P10–1a, construct the required cut and fill along the road. Use different hatching patterns for cut and fill. Plot at 1 in. = 50 ft on an 11 × 17-in. sheet of paper, and submit it to your instructor.

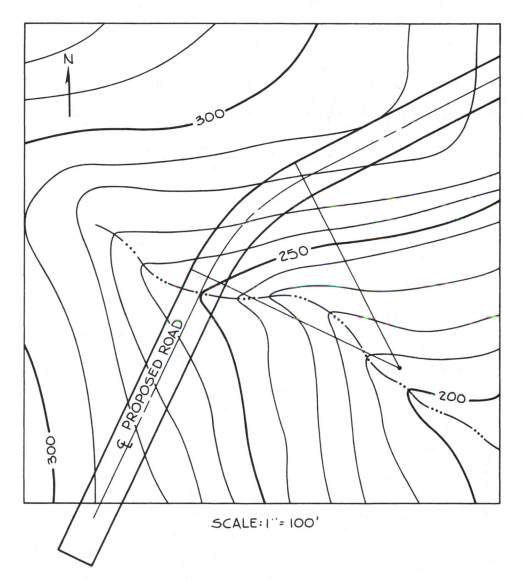

FIGURE **P10–1.**

P10–2 a. This exercise requires you to plot the cut and fill for an inclined road. You are given the plan, Figure P10–2, and must construct a profile from which to calculate cut and fill. Refer to Figure 10–4 if you encounter problems. The road is to have a 6 percent grade and the cut and fill begins at elevation 1500 ft. The angle of repose for both cut and fill is 2 : 1. Use shading and labels to identify the cut and fill in the plan view.

b. Transfer the graphics shown in Figure P10–2 to a CAD system, either by digitizing or by scanning. Follow the directions given in P10–2a. Plot at 1 in. = 50 ft on an 11 × 17-in. sheet of paper, and submit it to your instructor.

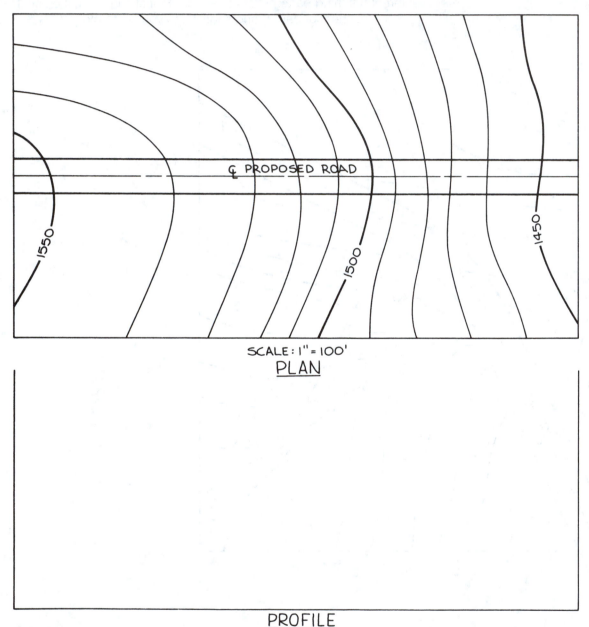

SCALE: 1" = 100'
PLAN

PROFILE

FIGURE P10–2.

P10–3 a. The cut and fill for a proposed industrial site must be determined in this problem. The site is shown in Figure P10–3. The elevation of the site is to be 640 ft. Angle of repose for cut is 2 : 1 and 1½ : 1 for fill. Plot the areas of cut and fill directly on the map and label them.

b. Transfer the graphics shown in Figure P10–3 to a CAD system, either by digitizing or by scanning. Follow the directions given in P10–3a. Use CAD software, if available, to construct the areas of cut and fill. Plot at 1 in. = 100 ft on an 8½ × 11-in. sheet of paper, and submit it to your instructor.

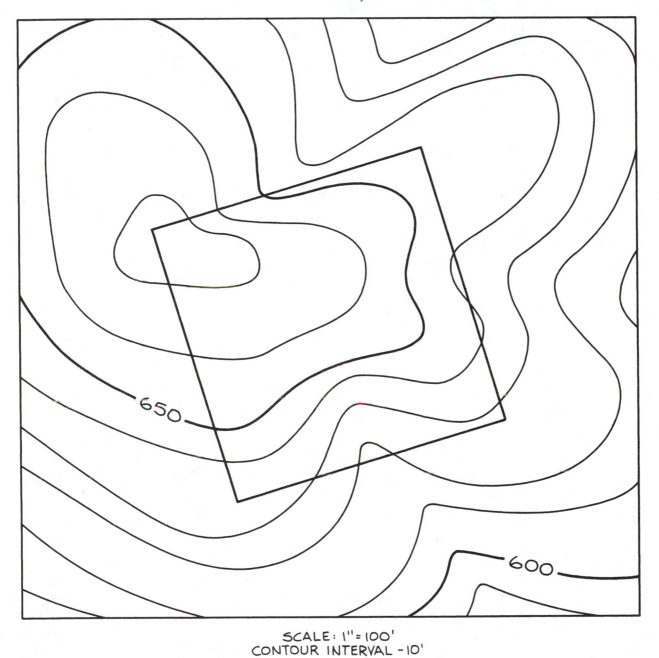

SCALE: 1" = 100'
CONTOUR INTERVAL – 10'

FIGURE **P10–3.**

P10–4 The measurements on the next page were surveyed on a grid for a borrow pit. The first column in the table represents the grid location. The "Existing Elevation" is the elevation before excavation, and the "Finish Elevation" is after excavation. The grid spacing is 50 ft. Grid lines A–E run north and south. Draw the borrow pit grid to scale on a B-size sheet. Label the grid values, and label the cut depths at each grid intersection. Calculate the total amount of excavated material in cubic yards. Place this number on your drawing.

Grid	Existing Elevation	Finish Elevation	Grid	Existing Elevation	Finish Elevation
A1	236.2	233.1	D1	238.3	231.5
A2	235.8	232.8	D2	237.6	232.2
A3	237.0	233.1	D3	236.9	231.8
A4	236.9	234.5	D4	239.2	234.1
A5	237.1	233.9	D5	240.2	235.4
A6	238.2	234.1	D6	238.7	234.1
B1	237.1	233.0	E1	238.3	234.0
B2	238.2	234.0	E2	239.2	235.6
B3	239.4	235.3	E3	237.5	233.2
B4	238.7	233.9	E4	239.7	234.2
B5	239.6	235.2	E5	240.8	235.2
B6	240.1	236.4	E6	241.3	235.9
C1	240.8	235.4	F1	241.3	236.5
C2	239.8	234.5	F2	240.6	236.1
C3	239.5	234.2	F3	239.8	235.3
C4	240.3	236.1	F4	240.7	236.2
C5	239.1	234.8	F5	241.3	237.8
C6	240.6	236.0	F6	240.6	235.9

P10–5 The following measurements were surveyed on a grid for a borrow pit. The grid spacing is 20 ft. Draw the borrow pit grid to scale on a B-size sheet. Label the grid values, and label the cut depths at each grid intersection. Calculate the total amount of excavated material in cubic yards. Place this number on your drawing.

Grid	Existing Elevation	Finish Elevation	Grid	Existing Elevation	Finish Elevation
A1	547.3	543.1	D1	546.3	542.8
A2	546.7	542.6	D2	546.9	543.1
A3	546.8	541.9	D3	547.3	544.0
A4	546.3	541.5	D4	547.6	543.8
A5	547.4	542.8	D5	547.9	544.2
A6	547.7	543.1	D6	547.5	543.8
B1	548.3	544.2	E1	548.4	544.3
B2	548.5	544.8	E2	548.2	543.7
B3	548.9	544.6	E3	548.0	544.3
B4	548.2	543.9	E4	547.8	543.4
B5	547.4	543.7	E5	547.9	544.0
B6	547.8	543.6	E6	547.5	543.2
C1	547.3	543.1	F1	547.8	543.1
C2	547.0	542.8	F2	548.2	543.9
C3	547.0	542.5	F3	548.5	544.6
C4	546.8	542.3	F4	548.3	544.0
C5	546.4	542.0	F5	548.8	544.7
C6	546.1	542.1	F6	548.3	544.8

P10–6 Plot cross sections for each of the profiles listed in the field notes in Figure 10–6. Use a single D-size sheet, or two C-size sheets for the layout. In each cross section, plot a 24-ft-wide level road with 4-ft-wide shoulders. Road elevation is 875 ft. Angle of repose for fill is 1½ : 1 and 1 : 1 for cut.

Label the road centerline and dimension road and shoulder widths.

The road should slope 0.04 ft/ft from center line to edge of road. Shoulder slope is 0.06 ft/ft. Label these values on your drawing.

Draw the cut-and-fill beginning at the outside edge of the road shoulder. Label cut and fill with the appropriate symbol and values.

Draw the existing grade as a dashed line in the areas of cut and fill.

Use the earth sectioning symbol in areas of undisturbed earth.

P10–7 Plot cross sections for each of the profiles listed in the field notes in Figure 10–6. Use a single D-size sheet, or two C-size sheets for the layout. In each cross section, plot a 24-ft-wide level road with 4-ft-wide shoulders. Road elevation is 872 ft. Angle of repose for cut is 1:1. After drawing each cross section, determine the elevation at the intersection of the grade line and cut line. Use these elevations to calculate the earthwork involved between stations 5 + 00 and 7 + 00. Use the average end method illustrated in Figure 10–14 to calculate volumes of material in cubic yards. Record your answers below or on the drawing by each cross section.

Station   *Area*

5 + 00   _____

5 + 50   _____

6 + 00   _____

6 + 50   _____

7 + 00   _____

             Cubic yards = _____

CHAPTER 11

Introduction to Geographic Information Systems (GIS)

This chapter introduces and describes a geographic information system (GIS). It further presents information about GIS applications, training, and trends.

The topics covered include:

- Introduction to GIS
- GIS concepts
- GIS components
- Data formats
- Related disciplines
- GIS applications
- GIS industry
- Trends in GIS

INTRODUCTION TO GIS

The National Center for Geographic Information and Analysis (NCGIA) describes a GIS as "a computerized database management system for capture, storage, retrieval, analysis, and display of spatial data." Figure 11–1 shows how a GIS stores multiple layers of reality.

GIS software provides a set of tools for solving spatial problems by working with spatial and attribute data. *Spatial data* describes a feature's location, and *attribute data* describes the characteristics of the feature. This is shown in Figure 11–2.

GIS is the geographical equivalent of a spreadsheet because it provides answers to "what if" questions that have spatial dimen-

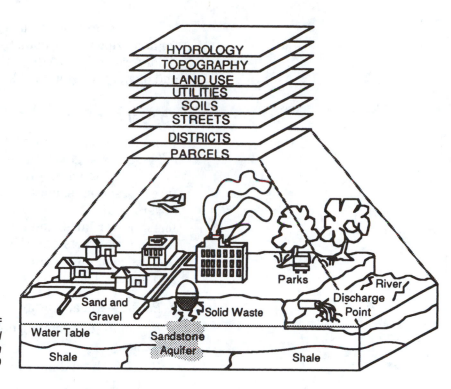

FIGURE 11–1. The real world consists of many geographies. *(Graphic image supplied courtesy of Environmental Systems Research Institute, Inc.)*

sions. Although many other computer programs can use spatial data (such as AutoCAD and statistical packages), only GIS programs have the ability to perform spatial operations.

The technology allows workers to function more efficiently and effectively. GIS software provides better decisions based on better information.

GIS CONCEPTS

Geographic information has two related parts: spatial and attribute data. The spatial and attribute data are related by a common item that has a unique value. This relationship is called the *georelational data model.*

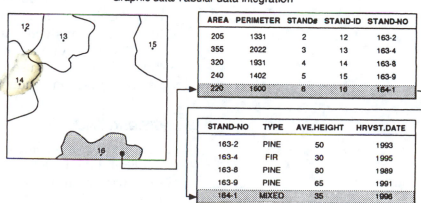

FIGURE 11–2. Spatial and attribute data related by a unique identifier (STAND-NO). *(Graphic image supplied courtesy of Environmental Systems Research Institute, Inc.)*

Graphic data/Tabular data integration

AREA	PERIMETER	STAND#	STAND-ID	STAND-NO
205	1331	2	12	163-2
355	2022	3	13	163-4
320	1931	4	14	163-8
240	1402	5	15	163-9
220	1600	6	16	164-1

STAND-NO	TYPE	AVE.HEIGHT	HRVST.DATE
163-2	PINE	50	1993
163-4	FIR	30	1995
163-8	PINE	80	1989
163-9	PINE	65	1991
164-1	MIXED	35	1996

The georelational data model offers a consistent framework for analyzing spatial data. By putting maps and other kinds of spatial information into digital form, GIS allows users to manipulate and display geographical knowledge in new and exciting ways.

Spatial Data

From 80 to 90 percent of the information that utilities, planners, engineers, and governments use relates to locations on the earth.

GIS programs store spatial data as one of three primary feature types: points, lines, and polygons. Power poles, maintenance covers, and wells are examples of point features. Street centerlines, power lines, and contours are line features. Parcels, soils, and census tracts are polygon features. Other feature types that are designed for specific GIS applications include: routes, regions, and voxels. These features can be briefly defined:

- *Routes*—linear features used by the transportation industry.
- *Regions*—overlapping polygons used by the mineral extraction industry.
- *Voxels*—three-dimensional polygons used by model flow of air and water.

A GIS makes connections between activities based on location. This connection can suggest new insights and explanations. For example, we can link toxic waste records with school locations through geographic proximity.

Attribute Data

GIS programs use *relational database management systems* (RDBMS). The RDBMS relates information from different files. Each related file contains one data item that is the same.

Topology

Topology is a branch of mathematics that deals with relationships among geometric objects. Topology defines and manages relationships such as connectivity, adjacency, and contiguity. These relationships are intuitively obvious to humans but difficult for the computer to determine.

Topology explicitly defines boundaries for the computer. Further, topology is used to identify editing errors, perform polygon overlays, and conduct network analysis.

GIS and CAD

GIS and CAD are complementary technologies. Both systems use a digital data model to enter, store, query, and display information; with both, the user interacts via screen menus and commands.

Comparisons Between GIS and CAD

- GIS data is in a database; CAD data is in the graphics.
- GIS data structures make use of topology; CAD data structures do not make use of topology.
- GIS coordinates are georeferenced; CAD coordinates are geometrically referenced.
- GIS databases have mapping accuracy; CAD drawings have engineering accuracy.
- GIS represents the world as it exists; CAD designs and drafts human-made and natural objects.

GIS COMPONENTS

Data Analysis

Spatial analysis permits synthesis and data display in new and creative ways, using spatial and attribute queries.

    Spatial queries act on points, lines, and polygons. The following are examples:

- *Point-in-polygon* (shown in Figure 11–3): Locates points that are within polygons.

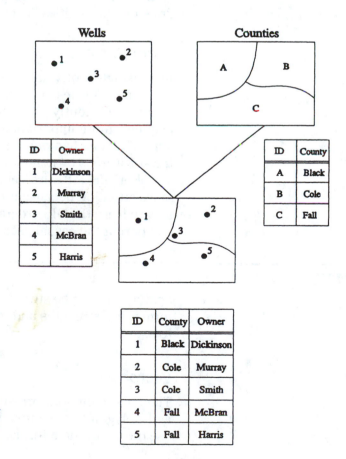

FIGURE 11–3. Point in polygon is an example of a spatial analysis. *(Courtesy of NCGIA, University of California at Santa Barbara.)*

- *Polygon overlay:* Generates new polygons from two or more existing polygons.
- *Buffering:* Creates new polygons around a set of points, lines, or other polygons.
- *Nearest-neighbor:* Identifies features that are closest to another feature.
- *Network analysis:* Performs flow analysis, determines routing, or orders stops.

Attribute queries act on the features' attributes. The following are examples:

- *Extracting:* Reselects a subset from existing features.
- *Generalizing:* Combines polygon features based on similar attributes.

Data Display

Maps communicate geographic information in a clearly understood and easily interpreted format. GIS maps allow relationships between different types of data to be clearly seen.

GIS programs do not have any cartographic intelligence that can guide operators in the choice of map symbols and other graphic effects. The programs have tools to symbolize maps and draw neatlines, titles, legends, text, and scale bars.

User Interface

Increasingly, software developers are using windows-like environments. The design of these environments determines how easily and efficiently the end users perform tasks. This design becomes more important as software vendors develop new markets—such as banking and real estate—where the operation of the software is by non-GIS employees.

For GIS software developers, the user interface is receiving more attention because it may reduce the amount of time necessary to learn the GIS program, provide quicker access to the data, and permit the casual user to make maps easily.

DATA FORMATS

Spatial data can be stored in two formats, *vector* and *raster*. Each format has strengths and environments where it is advantageous to use it.

Vector

The most common format for storing spatial data (points, lines, and polygons) is vector. All computer-aided drafting programs use the vector data format. Figure 11–4 shows the vector format.

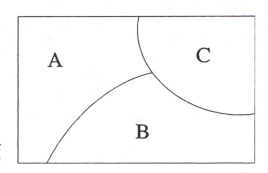

Vector formats work best with data that have a well defined boundary: parcels, street right-of-ways, political boundaries, and other human-made artificial boundaries.

The vector format provides greater accuracy and resolution than raster. Vector can also store data in a minimum amount of disk space.

Raster

The raster format consists of an array of cells, sometimes called *grids* or *pixels* (*pic*ture *el*ements). Each cell references a row and column, and contains a number representing the type or value of the attribute being mapped. Figure 11–5 shows the raster format.

Raster formats work best with features that are continuous and that change their characteristics gradually over distance— such as soils, vegetation, and wildlife habitat.

Scanned images and remote sensing data, including aerial and space photographs and satellite imagery, use the raster format. Electrostatic plotters produce maps in a raster format, at a resolution that makes the lines appear smooth.

Spatial analysis is easier to perform with a raster format. Unlike the vector format (which calculates line intersections), the raster system evaluates the values contained within each overlying cell.

Two principal disadvantages of raster formats are the large amounts of storage space required to hold the data and the lack of accuracy when using large scale maps.

A	A	A	A	C	C	C	C
A	A	A	A	C	C	C	C
A	A	A	B	B	C	C	C
A	A	B	B	B	B	B	B
A	B	B	B	B	B	B	B

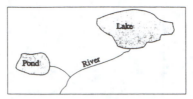

Reality - Hydrography

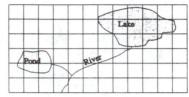

Reality overlaid with a grid

0	0	0	0	0	0	0	1	1	1	0	0
0	0	0	0	0	0	1	1	1	1	1	0
0	0	0	0	0	0	0	1	1	1	0	0
0	1	1	0	0	0	2	2	0	0	0	0
0	1	1	2	2	2	0	0	0	0	0	0
0	0	0	2	0	0	0	0	0	0	0	0

0 = No Water Feature
1 = Water Body
2 = River

Resulting raster

FIGURE 11–6. Creating a raster GIS from a vector GIS. *(Courtesy of NCGIA, University of California at Santa Barbara.)*

Format Conversions

Rasterization is the conversion of vector data to raster. This process is relatively easy and is shown in Figure 11–6. *Vectorization* is converting raster data to vector. This process is time-consuming and prone to errors.

As scanning and remote sensing become popular methods of building and updating GIS databases, GIS software packages have included the capability to convert between raster and vector formats.

RELATED DISCIPLINES

GIS is an "enabling technology" because of the assistance it offers to the many disciplines that use spatial data. A GIS brings these disciplines together by emphasizing analysis, modeling, and integration. GIS often claims to be the science of spatial information, and functions as the tool that integrates a variety of data.

Each of the following disciplines provides some of the techniques that comprise GIS.

- *Geography:* Provides techniques for spatial analysis.
- *Cartography:* Focuses map design and data display.
- *Remote sensing:* Provides image analysis and data input.

- *Photogrammetry:* Calculates measurements from aerial photos.
- *Civil engineering:* Applies GIS in transportation and urban design.
- *Geodesy:* Furnishes highly accurate positional control for data.
- *Surveying:* Supplies positional data on natural and human-made objects.
- *Statistics:* Provides techniques for data analysis.
- *Mathematics:* Provides geometry and design theory.
- *Computer science:* Provides software, hardware, and techniques for data entry, display, and management.

GIS APPLICATIONS

Parcel-Based

Land use planning is the most common GIS application, because of GIS' ability to perform "what-if" scenarios.

Some other parcel-based applications include processing building permits, assessing taxes, inventorying vacant land, and adjusting school enrollment. Figure 11–7 shows a parcel base map.

FIGURE 11–7. Tax lots and street right-of-ways in a parcel base map. *(Reproduced by permission of Metro, Portland, Oregon.)*

Natural-Resource-Based

Some of the first GIS applications were in forestry. Private timber companies use GIS programs to assist with planning and managing timber ownership, to plan harvest rotations, and to inventory timber and soil types.

Other GIS applications include identifying wildlife habitats, analyzing environmental impacts, estimating earthquake impacts, and monitoring surface water pollution.

The oil and gas industries use GIS extensively to identify and develop new prospects for exploration. They also use GIS to manage their current lease holdings and pipeline and production facilities.

Civil-Engineering-Based

Civil engineers use GIS to provide a spatial context for efficiently managing our infrastructure, such as water and sewer systems, bridges, roads, airports, and solid waste facilities. GIS assists engineers in siting public and private developments by providing information about the potential impacts of the proposed development.

Some other engineering and public work applications include monitoring pavement conditions, tracking street lighting, siting landfills, and designing transit corridors.

Demographic Analysis

The 1990 Census provides a wealth of geographically referenced socioeconomic and demographic data. Governments use this data to support the provision of services. Businesses use census data in daily decision-making processes. GIS programs can use census data to conduct demographic analysis.

Using GIS to redistrict is an example of demographic analysis. Other examples include forecasting population and employment, identifying crime patterns, analyzing retail market potential, and siting branch locations.

Street-Network-Based

In preparation for the 1990 Census, the Bureau of the Census developed a major national database called the Topological Integrated Geographic Encoding and Referencing (TIGER) Files.

TIGER is the first comprehensive digital street map of the United States. TIGER's features include all streets and railroads, significant hydrographic features, Native American reservations, military bases, and political boundaries.

Choices in Waste Flow
Solid Waste Simulation Model

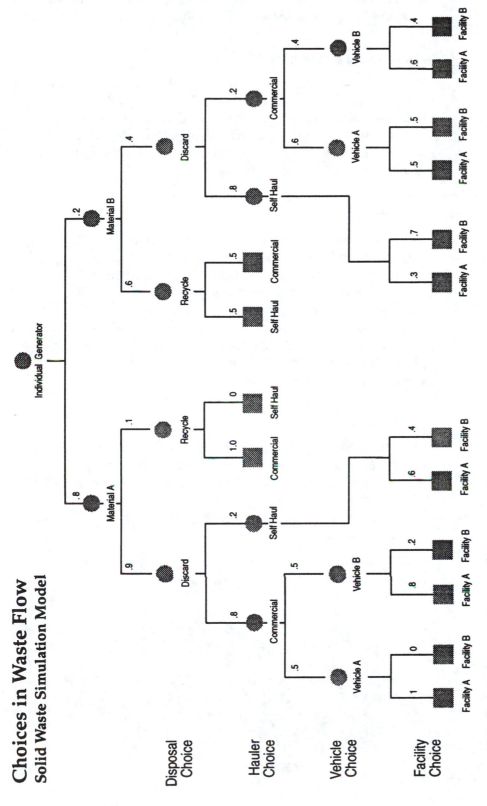

FIGURE 11–8. Diagram of solid waste flow model. *(Reproduced by permission of Metro, Portland, Oregon.)*

Some street-based applications that use GIS technology include planning new highways, routing local delivery services, and scheduling long-distance freight haulers.

Facilities Management

The utilities—water, waste water, gas, electric, and telecommunications—have been leaders in using GIS (called Automated Mapping and Facilities Management AM/FM systems).

Examples of facilities management include siting transmission facilities, relicensing of hydroelectric projects, routing meter readers, and tracking energy use.

Modeling

One of the most powerful uses of GIS is modeling. Current applications include modeling solid waste flow, projecting urban growth patterns, and anticipating the spread of oil spills.

GIS complements other computer-based models that deal with georeferenced data such as economics, transportation, deforestation, and agricultural productivity. GIS provides input data for the model and then displays the model's output. An example is the solid waste flow model used by Metro in Portland, Oregon, shown in Figure 11–8.

3D Analysis

Identifying all locations that can be seen from a specific site is called *viewshed determination*. This is difficult with traditional map analysis methods but relatively easy and accurate with GIS. Viewing proposed impacts, such as tree harvesting or commercial development, helps avoid unsightly mistakes. Figure 11–9 shows the 3D perspective.

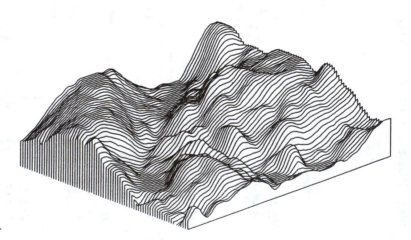

FIGURE 11–9. A 3-dimensional GIS.

History

In 1960, Canada became the first nation to develop a national GIS database. This GIS database gave Canada the ability to conduct nationwide geographical analysis. The New York Department of Natural Resources developed the first state GIS in 1975 around an inventory of their land use and vegetation. The next states to develop GIS programs were Minnesota, Maryland, and Texas.

During the 1960s, GIS programs focused on complex mathematics and were proprietary, expensive, and slow. During the 1970s, the focus shifted to data conversion and justification of the high cost of that conversion.

Software matured in the 1980s, and the performance of workstations and personal computers increased rapidly. This increase in price performance prompted an annual growth rate of over 35 percent, and an explosion in the number of GIS systems. These systems focused on integrating existing databases, data quality (or "fitness for use"), applications, and enhanced graphic capabilities.

Desktop mapping with GIS increased in the early 1990s. Software developers began to improve the user friendliness of GIS software.

Software/Hardware

GIS users are continuing to migrate from mainframes and mini-computers to workstations and personal computers. The increased power and reduced price of personal computers, coupled with the availability of larger mass storage devices (such as CDs and optical disks), allow users to meet their needs with a PC-based GIS. The shift to PC-based GIS increases the number of GIS users, because more organizations will be able to afford GIS technology.

Additionally, more GISs are being networked, and users are producing higher-quality color output.

Third Party Applications

The toolbox design of GIS software lends itself to specialized third party applications. The number and diversity of these applications increase as divergent businesses adopt GIS technology.

Third party applications include exploring for natural resources, dispatching emergency services, and evaluating the impact of global warming. GIS may either provide the input data or serve as the user-friendly front end for these applications.

Training

Due to the dynamic nature of GIS, ongoing training and education are essential. While training on a specific GIS software is invaluable, a good understanding of GIS concepts is equally important.

Two primary education tracks train users. One is for those who want to develop the technology itself, such as computer programmers. The second is for those who want to apply GIS within a particular discipline, such as soil scientists.

TRENDS IN GIS

Computers make mapping possible for those who have no formal training in cartography but who work with geographical data. Cartographers have valid concerns about the quality of output that these users produce. On the other hand, more people are able to use GIS; in and of itself this will increase the acceptance of GIS.

Desktop Mapping

As the GIS industry matures, desktop mapping is a natural evolution. *Desktop mapping* software allows a PC-based GIS to be easier to use than mainstream GIS software. Desktop mapping has a graphical user interface (GUI) that conforms to the Microsoft Windows standard. A *graphical user interface* is the manner in which information and options are displayed for you by the software. One of the most common GUIs in Microsoft Windows is the dialog box. A *dialog box* is accessed by a command or option in the software. The dialog box may contain a variety of information that you pick for use. The use of dialog boxes saves time and increases productivity by reducing the amount of typing you must do. The GUI provides the casual GIS user access to spatial databases and the ability to easily produce maps, charts, and graphs. Desktop mapping has quickly become the most common GIS system in use today.

Although the lines between mainstream GIS and desktop mapping may be blurred, several key features are found in most desktop mapping software packages. These features include:

- Reliance on the mouse as the primary method of interacting with the software.
- Ability to easily import maps, charts, graphs, and tables into other software packages.
- Quick access to spatial data.
- Simplified tools for display of spatial data.
- Large data packages bundled with the product.

Desktop mapping software comes bundled with several hundred megabytes of data. This data can encompass the entire United States with data layers such as counties, cities, road, rivers, and other distinguishing features. The purchaser generally has the option to purchase, at a minimal cost, additional data such as demographics, lifestyle data, traffic counts for major highways, and weather-related data.

Hypermedia

Hypermedia is a system to reference and retrieve different forms of digital data. It is becoming more integrated with GIS. Some examples of the digital data used by GIS include satellite images, architectural and engineering drawings, survey information, aerial photography, video, and sound.

Referencing digital data geographically allows the end user to access and analyze data in new ways. This analysis reveals new associations and reveals existing relationships.

TEST

Part I

11–1 Why is GIS considered an enabling technology?_____

_____

11–2 What is the most common GIS application?_____

_____

11–3 Considering all of the GIS systems in use today, what type of GIS system is the most common?_____

_____

11–4 Describe three types of spatial queries._____

_____

11–5 List four GIS applications._____

_____

11–6 List three disciplines that provide techniques used by GIS._____

_____

PART II

11–1 What happens to the data if it is converted from a raster format to a vector format?_____

_____

11–2 What is the term that determines the conversion of vector data to raster data?_____

_____

11–3 Describe how a raster GIS stores data. Give three examples of the type of features where the raster format works best._____

_____

_____

11–4 Describe two advantages of a vector GIS. Give three examples of the type of features where the vector format works best._____

_____

_____

11–5 List two advantages of a well-designed GIS interface._____

_____

_____

11–6 Fill in the blanks: _____ _____ permits synthesis and data display in new and creative ways, using spatial and attribute queries.

11–7 Fill in the blank: A _____ relates spatial data to attribute data.

11–8 What is the type of analysis that topology permits? _____

_____

11–9 Give four examples of spatial data used on a regular basis by civil engineers. _____

_____

11–10 What type of database management system is used by GIS software?

_____

11–11 Provide two reasons why GIS users are moving from mainframes and minicomputers to personal computers. _____

_____

11–12 Define *desktop mapping.* _____

_____

PROBLEMS

P11–1 Using your own words, describe three differences between CAD and GIS systems._____

_____

P11–2 Name three parcel-based GIS applications._____

_____

P11–3 Describe topology in your own words._____

_____

P11–4 Why would the raster format not be acceptable for storing parcel information?_____

_____

P11–5 Show a soil map in raster GIS format with at least three different soil types._____

_____

Glossary

Aeronautical Charts. special maps used as an aid to air travel.

Angle of repose. the slope of cut and fill from the road expressed in feet of horizontal run to feet of vertical rise.

ASCII. American National Standard Code for Information Interchange. Information contained in a computer file can be read by any computer.

Azimuth. a horizontal direction measured in degrees from 0 to 360. Usually measured from north.

Base line. a principal parallel used in establishing the rectangular system of land description.

Bearing. the direction of a line with respect to the quadrants of a compass, starting from north or south.

Bench mark. a reference or datum point from which surveys can begin. These are often brass caps mounted in concrete. Temporary bench marks may be iron pipes or spikes used for small surveys.

CAD (Computer-Aided Drafting). drafting accomplished with the use of a computer and one of any number of software packages that converts information given by the user into graphics.

Cadastral maps. large scale maps depicting features in a city or town.

Cartography. the art of making maps and charts.

Cesspool. a holding tank used to break down and distribute waste materials to an area of earth.

Chain. a measurement tool composed of links originally 66 feet in length. Steel tapes of 100 feet long are often referred to as chains.

Civil Engineering. a discipline concerned with the planning of bridges, roads, dams, canals, pipelines, and various municipal projects.

Closed traverse. a survey in which the lines close on the point of beginning or on another control point. Most often used for land surveys.

COGO (Coordinate Geometry). a method of creating drawings with CAD that utilizes points or coordinates, along with geometric angles, and survey lengths and bearings.

Connecting traverse. a survey which closes on a datum or control point other than the point of beginning.

Construction survey. a localized survey in which building lines, elevations of fills, excavations, foundations, and floors are established and checked.

Contour interval. the vertical elevation difference between contour lines.

Contour line. a line used to connect points of equal elevation.

Control point survey. the surveyor determines the elevation of important points in a plot of land. These points are plotted on a map and contour lines can then be drawn to connect them.

Cross section. a profile or section cut through the land to show the shape or relief of the ground.

Curve length. the length of a highway curve from beginning to end measured along the centerline.

Cut and fill. a road construction term that describes the quantities of earth removed from hillsides and filled into low spots.

Deflection angle. in surveying, an angle that veers to the right or left of a straight line, often the centerline of a highway, power line, etc.

Degree of curve. the angle of a chord (from the preceding one) that connects station points along the centerline of a highway curve.

Delta angle. the central or included angle of a highway curve.

Digitizing. a method of transferring data from paper-based drawings to CAD. It is done with a digitizer and a mouse or puck. The information is entered by picking points off the existing drawing with the puck.

Distance meter. a surveying instrument employing electronics (lasers, radar, etc.) to accurately measure distances. Often termed an EDM (electronic distance meter).

DTM (Digital Terrain Model). a map that shows elevation changes in land forms. The map appears three-dimensional, and is created with the use of a computer.

DXF (Drawing exchange format). this industry standard format is a way in which drawing information may be exchanged from one CAD software package or model to another.

Easting. the distance of a point east from an origin based on a State Plane Coordinate System, or on another coordinate grid system such as the Universal Transverse Mercator (UTM) grid.

Effluent. wastewater that leaves a septic tank.

Elevation. altitude or height above sea level.

Engineering maps. detailed maps of a construction project.

Engineer's scale. a tool used by the drafting technician to accurately measure distances on a map.

Equator. a line that circles the earth at 0° latitude.

Foresight. a rod location in surveying from which an elevation and/or location reading is taken. The foresight will become the next instrument set-up point because the survey moves in the direction of the foresight.

Geodetic survey. large areas of land are mapped and the curvature of the earth becomes a factor. The process of triangulation is used.

Geographical maps. small-scale maps depicting large areas on the earth.

GIS (Geographic Information System). a database or an inventory of many different types of information that can be accessed and combined according to particular needs. The inventory can include features such as topography, city streets, utility pipelines, transmission lines, and land use, among others.

Grade. an established elevation of the ground or of a road surface.

Graphic scale. a scale resembling a small ruler in the legend or margin of the map.

Grid survey. a plot of land is divided into a grid, and elevations are established at each grid intersection. Contour maps can be drawn from the grid survey field notes.

Hydrologic map. a map depicting boundaries of major river basins.

Interior angle. the angle between two sides of a closed or loop traverse measured inside the traverse. Also known as an included angle.

Interpolation. to insert missing values between numbers that are given; an educated guess.

Invert elevation. the bottom inside elevation of a pipe.

Land survey. a survey that locates property corners and boundary lines; usually a closed traverse.

Latitude. an angle measured from the point at the center of the earth. Imaginary lines that run parallel around the earth, east-west.

Legend. an area on the map that provides general information such as scale, title, and special symbols.

Level. a surveying instrument used to measure and transfer elevations. Occasionally used for distance measurements.

Local attraction. any local influence that causes the magnetic needle to deflect away from the magnetic meridian.

Longitude. imaginary lines that connect the north and south poles.

Lot and block. a method that describes land by referring to a recorded plat, lot number, county, and state.

Magnetic declination. the horizontal angle between the magnetic meridian and the true meridian.

Magnetic meridian. the meridian indicated by the needle of a magnetic compass.

Maps. graphic representations of part of or the entire earth's surface drawn to scale on a plane surface.

Map scale. aid in estimating distances.

Meridians. lines of longitude.

Metes and bounds. a method of describing and locating property by measurements from a known starting point.

Military maps. any map with information of military importance.

Mylar. the trade name of plastic media used as a base for drawings in the drafting industry. Also known as polyester film.

Nautical charts. special maps used as an aid to navigators.

Northing. the distance of a point north from an origin based on a State Plane Coordinate System, or on another coordinate grid system such as the Universal Transverse Mercator (UTM) grid.

Numerical scale. the proportion between the length of a line on a map and the corresponding length on the earth's surface.

Open traverse. a survey that does not return to the point of beginning and does not have to end on a control point.

Photogrammetric survey. aerial photographs taken in several overlapping flights. A photogrammetric survey becomes the "field notes" from which maps can be created.

Plan and profile. a drawing composed of a plan view and profile view (usually located directly below the plan). This type of drawing is often created for projects such as highways, sewer and water lines, street improvements, etc.

Plat. a map of a piece of land.

Plot plan. similar to a plat but showing all buildings, roads, and utilities.

Plumb bob. a pointed weight with a line attached to the top used in locating surveying instruments directly over a point or station. In chaining, used to locate exact distance measurements directly over a station point.

Point of curve. the point at which a highway curve begins.

Polygon. a multisided figure. If the included angles of a four-sided polygon equal 360°, the polygon will close.

Principal meridian. a meridian used as a basis for establishing a reference line for the origin of the rectangular system.

Profile. an outline of a cross section of the earth.

Quadrant. the compass circle is divided into four 90° quadrants: northeast, northwest, southeast, and southwest.

Radius curve. the radius (measured in feet or meters) of a highway curve.

Rectangular system. a system devised by the U.S. Bureau of Land Management for describing land in terms of blocks in a state plane coordinate system.

Relief. variations in the shape and elevation of the land. Hills and valleys shown on a map constitute "local relief."

Representative fraction. distance on map/distance on earth.

Rod. a square-shaped pole graduated to hundredths of a foot used to measure elevations and distances when viewed through a level or transit.

Route survey. an open traverse used to map linear features such as highways, pipelines, and power lines. A route survey does not have to close on itself or end on a control point.

Saddle. a low spot between two hills or mountain peaks.

Scanner. a device that allows information to be transferred from paper-based drawings to CAD drawings. It is moved over the existing drawing and "reads" two-dimensional lines, converting them to data that can be read by the CAD system.

Section. townships are divided into 36 squares—each square is a section. A section is one mile square containing 640 acres.

Septic tank. a concrete or steel tank used as a method of sewage disposal that disperses wastewater to a system of underground lines and into the earth.

SPC (State Plane Coordinate System). a grid system that has known and precise coordinates in relation to our three-dimensional earth. It allows portions of the earth to be drawn accurately in two dimensions.

Stadia. a type of distance measurement employing a level and a rod. Also a Greek term referring to a unit of length equal to 606 feet, 9 inches.

Station. arbitrary points established in a survey usually located 100 feet apart. An instrument set-up point is often referred to as a station.

Stereoscope. a device that allows the viewer to see elevations in aerial photographs by viewing two identical photographs side by side.

Subdivision. a parcel of land divided into small plats usually used for building sites.

Surveyor's compass. used in mapping to calculate the direction of a line.

Terrain. the shape and lay of the land.

Theodolite. a precise surveying instrument used to measure angles, distances, and elevations.

Topographic map. a map that represents the surface features of a region.

Topographic survey. a survey that locates and describes features on the land, both natural and artificial. Often accomplished through the use of aerial photography.

Topography. the science of representing surface features of a region on maps and charts.

Townships. an arrangement of rows of blocks made up of parallels and meridians. A township is six miles square.

Transit. a surveying instrument for measuring angles, elevations, and distance.

Transit line. the centerline of a linear survey (highway, pipeline, etc.).

Traverse. a series of continuous lines connecting points called traverse stations or station points. The surveyor measures the angles and distances of these lines, and these field notes are then transferred to a map or plat.

Triangulation. a series of intersecting triangles established as a reference in geodetic surveys. Some sides of these triangles may be hundreds of miles long and cross political boundaries.

Turning point. a temporary bench mark (often a long screwdriver, stone, or anything stable) used as a pivot for a rod. The turning point can be both a backsight and a foresight for the rod.

Utilities. service items to a home, business, or industry such as electrical, gas, phone, or TV cable.

Vellum. transparent paper used in the drafting industry.

Verbal scale. expressed in the number of inches to the mile.

Vertical curve. the shape of a linear feature such as a road or highway (in profile) as it crests a hill or creates a sag in a valley or depression.

Abbreviations

AB	Anchor bolt	CCP	Concrete cylinder pipe
ABDN	Abandon	CFM	Cubic feet per minute
ABV	Above	CFS	Cubic feet per second
AC	Asbestos cement, Asphaltic concrete	CI	Cast iron
ACI	American Concrete Institute	CISP	Cast iron soil pipe
ADJ	Adjacent, Adjustable	CJ	Construction joint
AHR	Anchor	CLG	Ceiling
AISC	American Institute of Steel Construction	CLR	Clear, Clearance
ANSI	American National Standards Institute	CMP	Corrugated metal pipe
APPROX	Approximate	CMU	Conc. masonry units
ASPH	Asphalt	CO	Cleanout
B & S	Bell and Spigot	COL	Column
BETW	Between	CONC	Concrete
BKGD	Background	CONN	Connection
BL	Base line	CONT	Continue, Continuous
BLDG	Building	CTR	Center
BLT	Bolt	CYL	Cylinder
BLW	Below	CONST	Construction
BM	Beam, Bench mark	D	Degree of curve
BOT	Bottom	D or DR	Drain
BRG	Bearing	DIA	Diameter
BSMT	Basement	DIAG	Diagonal
B/U	Built up	DIR	Direction
BV	Butterfly valve	DIST	Distance
BVC	Begin vertical curve	DN	Down
C TO C	Center to center	DWG	Drawing
CB	Catch basin	EA	Each

EDM	Electronic distance meter	MIN	Minimum
EF	Each face	MSL	Mean sea level
EL or ELEV	Elevation	MH	Manhole
EQL SP	Equally spaced	MIN	Minimum
EQPT	Equipment	MISC	Miscellaneous
EW	Each way	NA	Not applicable
EXP	Expansion	NTS	Not to scale
EXP JT	Expansion joint	O TO O	Out to out
EXST	Existing	OC	On center
EXT	Exterior, Extension	OF	Outside face
EVC	End vertical curve	OPNG	Opening
FCO	Floor cleanout	OPP	Opposite
FD	Floor drain	ORIG	Original
FDN	Foundation	OD	Outside diameter
FG	Finish grade	PC	Point of curve
FL	Floor, Floor line, Flow	PL	Plate, Property line
FLL	Flow line	PLG	Piling
FOC	Face of concrete	POB	Point of beginning
FPM	Feet per minute	PP	Piping
FPS	Feet per second	PRC	Point of reverse curve
FTG	Footing	PRCST	Precast
GPM	Gallon per minute	PRV	Pressure reducing valve
GPS	Gallon per second	PSIG	Pounds per square inch, gauge
GR	Grade	PT	Point of tangency
GVL	Gravel	PVC	Polyvinyl chloride
GND	Ground	PVMT	Pavement
GTV	Gate valve	PRESS	Pressure
HB	Hose bibb	R	Radius curve
HD	Hub drain	RC	Reinforced concrete
HDR	Header	RCP	Reinforced concrete pipe
HGT	Height	RD	Rain drain, Roof drain
HI	Height of instrument	REINF	Reinforce
HORIZ	Horizontal	REPL	Replace
ID	Inside diameter	REQD	Required
IE	Invert elevation	RMV	Remove
IF	Inside face	RW	Right of way
IN	Inch	SCHED	Schedule
INTR	Interior	SECT	Section
INVT	Invert	SH	Sheet
INFL	Influent	SPEC	Specification
INSTL	Installation	SQ	Square
JT	Joint	STL	Steel
L	Length of curve	STR	Straight
LONG	Longitudinal	STRUCT	Structure
LATL	Lateral	SUBMG	Submerged
MATL	Material	SYMM	Symmetrical
MAX	Maximum	STA	Station

T & B	Top and bottom	TRANSV	Transverse
T & G	Tongue and groove	TST	Top of steel
TBM	Temporary bench mark	TW	Top of wall
TC	Top of concrete	TYP	Typical
TEMP	Temporary	VC	Vertical curve
TF	Top face	VERT	Vertical
THK	Thick	W/	With
THKNS	Thickness	W/O	Without
TO	Top of	WP	Working point
TP	Turning point	WS	Water surface, Waterstop, Welded steel

Index

Volcanic activity, measuring of, 60
Voxels, 262

Water features in mapping, 121, 123, 124
West longitude, 73
West magnetic azimuth (WMA), 80

West magnetic declination (WMA), 80
Williamette meridian, 139, 141

Zenith angle, 50
Zero designation, 75
Zoning boundaries, 15

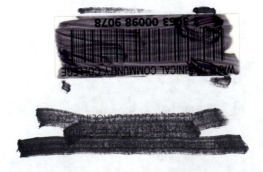